JESUS IS THE BEAM

Book Φne

Gwynevere Lamb

BookLocker

St. Petersburg, Florida

ISBN: 978-1-64718-554-1

Published by BookLocker.com, Inc., St. Petersburg, Florida.

Printed on acid-free paper.

BookLocker.com, Inc.
2021

Library of Congress Cataloging in Publication Data
Lamb, Gwynevere
Jesus is the Beam by Gwynevere Lamb
Library of Congress Control Number: 2020911312

This book is dedicated to those who reject God and put their faith in science. To those who doubt God and put their faith in themselves. To those who hate God and put their faith in no one. Humans are worshipers. We all worship something... rock stars, movie stars, or stars in the sky! Even if we claim to be totally non-religious, we are worshiping our own ideas.

This book is also dedicated to those who love God. To those who love God and science but have been frustrated by the scientific community's insistence that God and science are divorced, or their postulation that God and science were never married to begin with. Like me, I hope you are encouraged by God's insurmountable love and His beautiful creation as proof positive that there is a scientific explanation of God found in every good thing. After all, if God made it, He surely has a blueprint! I trust you agree (or will agree after reading this book) His triune nature of omniscience, omnipresence and omnipotence as well as the manifest trinity of God the Father, God the Son, and God the Holy Spirit are unified in one powerful force. I believe in the measurable mathematical evidence of God, and I love it with *all my mind*, but His immeasurable love, alone, should be enough reason for us to love him with *all our hearts, strength, and minds*.

Ruler of the Word

As you turn a page
From the book of your life
Sometimes you'll find
A tattered edge from strife.
Sometimes you'll find
A rip in the page
From many things
Trauma or rage.
Then there'r times
You'll find a page blank
You ponder and wonder
Did I turn the crank?

With each page you turn
You will always learn
The truth of your *yearn*!

Tragic chapters of defeat,
A special someone you did meet.
Page after page of contradictory
Or half a page of victory.
Pitfalls or blessings
Confidence or guessings
Or heart's broken in half the book

Of new love and a smile smitten
One thing's for sure
In the Book of Life
You want your name written!

Measure all footnotes with the Ruler of the Word!

2013

Table of Contents

Prologue

Writer's Block or Identity Mock

"This would be my thesis! Perfecto! I did have an edge," I convinced myself when I began to write this book 23 years ago. A miraculous event had taken place and I was running with fervor and so much zeal it was unthinkable at the time to consider it would be more than two decades before I published my story. I had a purpose, which was what fueled me and led me to stay on course.

When I would start to doubt, a rare circumstance would redirect me and align me with renewed confidence. In spite of the constant battle to esteem myself worthy of my task I had to remind myself that I am educated enough. Even though I felt dejected when MSU in Bozeman, MT rejected my application for graduate school, my under-graduate academic achievement was my only foundation. I would doubt myself again. Then God would remind me of 1 Corinthians 1:27. "God chose the foolish things of the world to shame the wise; God chose the weak things of the world to shame the strong."

Growing up I remember experts telling my parents that I was "slow" to learn. I was held back and had to retake the 2nd grade. I was not the sharpest pencil in the teacher's drawer, but I seemed to always have one in hand! I loved doodling! I loved to write and draw. In high school I was tutored by my childhood best friend, Mary, who was our class valedictorian, *every year*. My mother always said, *why can't you be more like Mary*. Yah, she was *that person...* my mother used her as a ruler for rearing me. I was never good enough or smart enough. Despite my lacking academic aptitude, by the time I reached my junior year I did very well in math classes and loved solving algorithms in algebra. I never got beyond that, though. I never took Calculus or Trigonometry, or Physics, not in school, anyway. My friend Mary took all those classes in AP courses, of course, and went on to more scholastic greatness, ending up in Yale's graduate program after graduating with honors from Denver University. All with full scholarships, no doubt! All I managed to achieve was entry into Western State (aka Wasted State) College (WSC) in Gunnison, CO, and I am still paying off student loans! For the record, Western State has come a long way and is now a state university! In spite of its past and long-standing mediocre reputation it was a great school and the faculty was a class act! What's more, attending WSC afforded me the bonus life-long lessons of unmatched outdoor education. I was a ski bum. Everyone at WSC was! Lol! Seriously, I can honestly say I am a true mountaineer! Learned it. Lived it! Priceless!

I maintained a steady closeness with my teachers throughout high school and college by volunteering as their aid. By doing so, I was able to learn and absorb information better and better with every new year. By the time I graduated college I maintained above average grades and I was honored a scholastic medal. Not a big deal to my Magna Cum Laude friends out there! It was a big deal, I told myself years later, recalling how I smoked pot every day, all day every day, even while taking exams.

I began to imagine how well I may have done had I been sober-minded!

After poor scores on the LSAT dismissed me from graduate school dreams, I was down and out until a couple years later when my life's purpose was poignantly revealed to me within a three-day period. "I will make scientific history!" I chimed to myself each time I was met with a closed door. I decided to write a book, and I did just that. I would integrate my life's story with science. I began with determination and passion until the day my computer died.

It crashed and I lost everything. Back then, in 1997, back-up drives were not mentioned. No one cautioned me to save stuff on floppy discs! I had to start over once I wiped away the tears and pulled myself up by my "ski-bootstraps." Painful as it was I did it. I began again, this time focused on the science and the math. I wrote well keeping everything in rule, but something was missing. The glue. Nothing was holding together. Fragments here and there. I was stuck. One day turned into one year, three years, more. I would go over what I had, but I just could not piece it together. There was no thrilling thread from one chapter to the next. The crude content was too raw to digest. I knew I needed to refine it to make it digestible, so I set it on a backburner. I turned it to simmer. Slow cooking.

Eventually the flame went out.

I would revisit it from time to time, stirring the pot. But the stew of my story was coagulated from coldness. Cold meat and potatoes in a cold gravy. Not very appetizing! The fire in my belly was simmering to glowing charcoal. My passion slowly began to fade into an unreachable realm. *So, I've got all this material, I've got all this knowledge, and I have it all backed up...* no chance in hell of losing it this time, I'd say. I had it but at a loss, still. It was as if I did lose it to hell, after all! What good was having it if I couldn't piece it together! I'd lift the lid to the pot of stew, take a peek and throw in a few ingredients. So, more years go by and nothing materializes, except an occasional carrot, potato, and sometimes a chunk of meat. *I'm doing it right this time,* I was sure! But then the hard truth of nothingness would poke me in the side and laugh at me. I began to fear that the stew of my story would remain on that back burner forever, and eventually rot.

I started over. Sort of. I tried a series of short stories in the first narrative. I thought if I go back and personalize it just a bit it would have more impact and substance to stick together. It was not enough. Over and over, I kept getting stuck and the frustration would cause me to quit. I would set it back on the backburner again. Then miraculously ten years later, a flame would ignite, and the stew would come to a boil, but, then a week or month later, nothing. The nutrients and flavor of my story began to disappear. My fervency began to follow. I was not consistent. Ten more years passed. The stew was turning to mush. How do you recover from mush? Most of us know what happens when we overcook meat and potatoes. The meat gets tender, yes, but what of the potatoes? That's right, MUSH. When you really overcook them the root and the starch break down and turn into a consistency that can't even be called *mushed* potatoes. It loses all cohesion and falls apart. Ick. And the meat now starts to break down into tasteless sinewy fibers. It's like eating stringed cardboard. Ew! So, this was the fate of my erudition. Yup. My ability, my mind and my book were headed toward a downward spiral of a black hole pulled by cosmic currents of a distant dimension to be sucked into the ocean of dark matter, never to be found again.

***

The whole reason my story remained on the back burner was because, ironically, *I would not come out of the closet.* I'd poke my head out once in a while. This would happen time and again.

The closet. Aahhh, the loathsome place from where or whence we all want to be free, right? In the closet I had to keep a smiling persona and play make-believe. I hid behind a mask and I wore shoes that were two-goody (or is it *too*-goody). I felt like I was under pressure to be the seminary-text-book definition of a "transformed born-again Christian." This made it entirely difficult when writing. I was afraid to be myself, much less share the truth about my shameful past. I was hiding behind other characters and narratives. I even convinced myself that if I wrote strictly in fiction I could preserve my identity. It wasn't until I fully came out of the closet, with its many dried bones and gave them to God, when I was able to write freely.

My story is unequivocally removed from that crockpot, and now I am more than relieved to begin afresh. I know how to run with it, and I am doing exactly that as I write this. And hey, it won't be so bad because I have saved fragmented pieces already written all over the place, right! Some pedantic, and some humorous, and some poetic. How harmonic it is to weave it all together. One minute you may find me astute with precision, measuring every little thing, and the next minute an unruly sarcastic court-jester-type saying things that would embarrass my grandmother, and next, a hopeless romantic erupting into poetic expression. One way or the other, I will articulate my story without fluff, off the cuff. No holds bar (whatever that means).

You're probably thinking, if you even made it this far, what is this rambling woman going on about? What the heck is this book about?

It's about the stuff I've found and the stuff that found me and founded me. My bruised and broken body in its journey with Jesus. The *unified body* and its healing and protective nature from maneaters and other eaters. This book's about a little bit of *everything*. Earthly adventures and heavenly science. The good and the bad... and the forces of nature from the boom of creation. This book eliminates the need for complicated calculus or trivial trigonometry as it focuses on the *writing on the wall* as it pulls the string from the entangled mess we've weaved. No fancy formulas here! No need for them as common sense and nature stitch the esprit fibers of our making into a glorious image crowning the evolutionary acme of the human race. In these pages we will reach the pinnacle of our sentient species, the force out to destroy us, and behold the soul and spirit within. We will unearth the key to unlock our molecular chains and atomic wires.

This book's about what I know from both sides of the wizard's curtain, and in 1997 I began to put everything on an atomic shelf and taxed it accordingly (*taxonomically*). All the pieces I carefully placed on the shelf, in their own compartments, are now joined together to reveal a complete picture, and stay fixed. God and science reunited! I have the glue. The truth. Amen! For a while there, I thought perhaps this would never get out and this wellspring of information would calcify inside me. I take no credit. This is all God, 100%.

I'm not hiding anymore! I had to remove my mask. I had so many. Literally, I had a mask collection. Lol! Joking aside, my husband, Aurie, and I enjoyed masquerade balls and dressing up for Halloween parties. *Being someone else is fun. What's the harm in it? Disguises and masks are alluring and add spice to life.* For me, I had so much spice it was burning a hole in the lining of my soul.

Opening the door to my skeleton closet was the most humbling thing I ever had to do. I kept it shut for so long because I was afraid of what others would think of me. I didn't want to be judged. God has now given me confidence to share my story, as is. He will not judge me because I've already judged myself as wretched and the worst among "men" and I've allowed Him full access to restore me which He is in the process of doing.

I'm not sure how well my style of writing will be received, but if I do say so myself, I find it delightfully refreshing and relieving, and that's only because I am writing now without any inhibitions. I'm finally free and I have lots of puns. Puns are fun! Tangents too. I gave myself the poetic license to be ridiculous. On a serious note, I will ask questions and make bold statements most people are too timid to mention or of which most are unaware. I will not be confined to a religiously acceptable box or popular culture's status quo, nor remain inside the artist's lines, or covered by the scientific blanket of explanations. To envision simple structures and simple systems with unobjectionable understanding we must remove all our coverlets. We must strip down to the bare minimum. In these pages I shed each layer, including my most beloved comforter of preconceived pieces which make up the quilt of my scientific and spiritual beliefs. Also, I've realized, that the scientific piece would not be as well-received without my personal journey to accompany it. And of course, the real me writing it… from start to finish. Well, not exactly… I love painting stories and in this book it brings to life the sojourns I've made not from the start and not finished yet, but enough in between to set the scene.

*It is my heart-felt plea in prayer this book anoints readers eyes and ears to hear God and see truth. I pray the blood of Jesus will break all hexes and spells of delusion and allow all who read this manuscript to be free from bewitching blindfolds and ears of distortion.

Disclaimer: Some language and tense of the language I use to paint past events does not necessarily reflect my current vernacular, present-day voice, or beliefs, including but not limited to expressions, agreements, and exclamations I use in dialogues and monologues. I simply stayed true to the story because this is my personal life's testimony. Some (not much) reader discretion is advised.

SHAME

Becoming more than one
To pacify the pain
To change one's face and name
The mind to entertain

Sustaining one's insanity
By playing in the game
By and large one thinks one's sane
To boot, full of fame

Plunging into pity
To satisfy the brain
To see oneself the same
Without blame or stain

Withdrawing from one's friends
By being what they call lame
By and by one feels vain
And has no sense of shame.

April 1, 2002

Chapter ΦNE

Who's Your Daddy
Circa 1985

It was late in the evening when the children were abruptly awakened by their mother. Her clothes disheveled, her hair a fright, her hand quaking but forceful, she shook them.

"Get up," she said with hardly a whisper, "now, be quiet! No need to change; put these on and let's go." She had their coats out and one by one, purposefully put them in it, one arm at a time times six.

She hurried the three children into the car and drove knowingly across town. Fifteen minutes later the car came to a stop and the engine turned off. They were in a parking lot of an unknown condo/townhome complex.

With the children once again fast asleep in the back, the mother turned around from the driver's seat to poke at their feet. She bit her lip, took a deep breath, and sighed. "Get up," she said this time more gently than when she first woke them from their beds. Her right arm outstretched; she nudged her daughter. "Wake up, we're here. Let's go," she said firmly, her face hidden in the shadows. "Wake your bothers."

She walked in large strides through the parking lot, down a sidewalk, into a corridor, up a flight of stairs, her kids in tow hurried to keep up. She stopped at a door and pulled out a set of keys to unlock it.

"Where are we, mommy?" They asked hesitantly, understanding that their questioning her was potentially going to anger her. A sharp look, and a tight upper lip silenced them as she held the door open for them.

The children stepped inside, curious, confused. To the left was a living room fully furnished with a luxurious sectional sofa, a fine coffee table with matching end tables. An entertainment center housing a big color television was center stage. To the right was a handsome dinette set, and a fully loaded kitchen. From left to right, were artwork and home interiors to boast about throughout.

"Take a look, children... look all this." Their mother stood there with one hand on her hip and the other hand waving across the two rooms. "This your daddy's house when he's not come home," she said with contempt in broken-English, her voice cracking. The three kids became wide-eyed and scattered to take a closer look at everything.

Straight ahead was a functional office which appeared to be in use, with its stacks of files and paper strewn about. Beyond that was a bathroom, and its counter was covered with make-up displayed as if it were for sale at a flea market.

"Look... all this make-up," their mother said with disgust. Her tone of voice told the children the make-up was not for her. If not hers, then whose? Was their father having an affair? No, of course not, he must be selling it, they thought.

"Where's daddy?" They asked. From the look on her face they could tell that their dad had no idea they were there. That explained why she was being so sneaky. She led them into the one bedroom lavished with fine linen and a bedroom set she could only dream of and pointed to the walk-in closet for the kids to enter. It was obvious she knew her way around, but it became abundantly clear that their father was clueless to her knowing about all this.

The children walked into the closet, in single file, their eyes wide and mouths agape to what they beheld. They couldn't believe their eyes. Sequined gowns, ruffle dresses, lacey nighties. Row after row. Costume jewelry spilling out of boxes spread out on four shelves, like found hidden treasure. Designer shoes, boots, pumps, stilettos, by the hundreds, so it seemed. All the glitter, all the sparkle; there was so much bling that the kids thought they walked into a dressing room at Disneyland.

They just couldn't believe their eyes. "Whose stuff is this," the three asked one after the other like an echo as they slowly touched everything as if it weren't real. The children couldn't decide whether to be excited or worried. On one hand, all this stuff told them that their dad must be rich, and on the other hand the stuff told them that there must be another woman. Looking up at his mom with puppy dog eyes for the sake of his father, knowing his dad got caught with his hand in the cookie jar, the youngest boy asked his mom, "is daddy dating a movie star?"

"No." Long pause. "Look up," and she pointed to the top shelf that ran across the entire closet. All three children turned in unison, like the three little kittens who lost their mittens. Afraid to lift their chins, as they were sad for their dad, they finally looked up, their heads tilting way back, they didn't understand what they were seeing. Columns and rows of boxes. Round boxes.

Their mother pulled one down, and grabbed another, and then another and another. In a frenzy she pulled out its contents. Hair! Wigs!

Blonde, brunette, red, long, short, curly, straight. In her madness, the children clung to each other and began to cry.

"Your father is dressing up as a woman! Can't you see! Look at these shoes, look how big!"

While my family grew to hate my dad, I grew to love him more. It cut me when I would hear my brothers call him faggot. It bruised me when I watched my mom scream in his face. I would have no part of it. To the contrary, I was very supportive of him, and once in high school my girlfriends and I attended his show. My dad was a female impersonator at a gay bar in old uptown Denver. *He was glamorous. He was fabulous!* My dad let us come over before the show and raid his closet. We all got dolled up together. My mom didn't know any of this. She would have surely condemned me and my dad to eternal damnation. From here on out I will refer to my mom as Oma, my translation for the Korean word for mom (pronounced um-mah), and the name I/we use to address her today. She was my mom, but it was my dad who expressed more maternal tenderness.

Sad to say, but I'm hard pressed to find a childhood memory of Oma embracing me and telling me she loved me. On the other hand, my dad would tuck me and my brothers in every night (when he was home) with a hug and a kiss, but not before he would tell one of his famous Chester stories. Chester and his dog Spot were always into some glorious adventure. Oh what fun we had listening to my dad make up these stories as he went along. What a talent he had. If only I could remember them, I would turn them into children's books. My dad, not surprisingly, took on the motherly role in our family. Unfortunately, his parental guidance was lost by his absence as we grew older. Yet, he remained the gentle spirit, full of x's and o's, while my mom held onto an iron rod with a clenched fist.

Weapon of Lass Destruction

My brothers and I were never allowed to spend the night at a friend's house because it was against Oma's rules. For my brothers, I don't think it was so bad because they had each other. They were only a year apart and so had a lot in common in the way of keeping each other company. I spent a lot of time with my dolls in my room. I have so many memories of being alone with only my toys and my imagination to keep me company. In middle school, at Ken Caryl Junior high School in Littleton, CO, the upcoming Sadie Hawkins Dance was all anyone could talk about. My friend Karen's parents were hosting a post-party for the event at their house, complete with snacks and refreshments, and a sleepover. Sleep-over! Impossible for me. So sad. But I could get in on some of the fun early on, I reasoned with myself. When I had mentioned the dance to Oma a month before the event, she without hesitating, told me I was not allowed to go. Of course, such events are cultivated for evil, and she must spare me from that sort of abomination.

So, when the dance neared to a week or so, I told my parents that there was an academic meeting I needed to attend in the evening at school, hoping they forgot about the dance I mentioned a month prior. I was able to go without a hitch. They didn't suspect a thing. I was having so much fun at Karen's post-party that I let the time get away from me. I told my parents that I would be home by 9 pm, and when it was nearly 10 pm I realized I was in trouble. I ran home as fast as I could. Five blocks later I was standing at the top of a very steep hill, divided in the middle by railroad ties. If it were any steeper I would call it a cliff. At the bottom of this 75 ft hill was my home. The railroad ties kept the hill from eroding and avalanching into our little cookie-cutter house. I could barely make it out as it was a dark night with only a third of a moon breaking in and out of the clouds. I flew down the hillside sidestepping. It's amazing I didn't lose my footing and fall, roll, and crash through the family room window. The house was as dark as I have ever remembered it. Every light was off, yes, but there was a foreboding darkness lurking at every corner that seemed to alarm me of what was in store for me. I snuck in the back door, tip-toed through the dining room and went directly to my room and into my bed, hoping everyone, including my parents, was asleep and didn't hear my entry.

Moments later my brother entered my room and told me that mom and dad were not home... they drove to the school looking for me. My heart beat out of my chest just thinking about what my parents would find once they got to my school. They'd find the cat out of the bag, for sure. Oma didn't take too kindly to lies, and I braced myself for the worst. The wait was torturous. I huddled in the corner of my room, in the dark, up against my bed, with my arms wrapped around my folded legs and my head buried into them. It was so quiet. I don't even remember hearing the garage door open or my parents coming in. All I recall while waiting was the sickening silence and the beating of my heart. I was terrified. My fears became reality when Oma stormed into my room with a horrible scowl on her face which was made even more horrific by the pale moonlight bouncing off of her as it barely shone through the window. She came at me like a bull and I was a matador waving a red cape. I took cover under my blanket. I will never forget the sting of every hit. I can't recall what weapon she used.

She favored wooden spoons when I was smaller, but at 12 years old, she would make me pick a switch from one of the bushes on the side of our house. If the switch was too flimsy for her liking she would go pick one herself and I would pay for the time it took her to do it with extra lashes to my backside.

The next morning, I woke to a sight I will never forget. My arms, legs, and back were covered in welts, some scabbed over from bleeding and others just puffed and bruised. This was by far, the worst "spanking" I had ever received. Oh, but more would follow, until the day I eventually ran away from home, more than three years later in 10th grade. Then it stopped. Had I known all I had to do was run away, I would have probably run away that night of the Sadie Hawkins Dance.

My brother, Solo (short for Solomon), doesn't remember getting belted much, so I guess the whippings were saved especially for me and my youngest brother, Zeke (short for Ezekiel). Solo was her favorite, no one would argue. And no one ever came to my or Zeke's rescue. Bitterness for Oma began to take root.

Chapter TWΦ

From Three to Two to Four

S cientists worldwide, for nearly a century, have been in a marathon... they have been experimenting, solving, digging, climbing, racing, crashing and smashing to find a grand unified field equation to explain... well uh, everything.

Everything!

Everything and everyone are included.

You and me,
Trees and bees,
Stars and snow,
Ticks and fleas

Atomic systems,
Systemic actions,

Cosmic rays and
Quantum fractions

The waxing moon,
The whining baby,
The deepest hole,
And *darkness*... maybe...

Physicists worldwide started racing in this marathon to unify everything since the early 20th century. The participants of this race have slowed their pace to walking because many believe they have found what they've been searching for, although a proper formula is yet to be seen. According to Stephen Hawking, in his book, A Brief History of Time, published in 1988, finding the proper equation to the theory of everything should happen by the year 2020.

It's a crazy rat race out there. The traffic. Most people in a rush to do nothing. To get home and watch television? While the mice run their race, physicists have been running theirs. In light of all the advancements of particle physics, it's somewhat hard to believe they're still stuck in the maze. They don't think they are, but from what I can tell, they are nowhere closer to finding the theory of everything than they were when Stephen Hawking made his prediction. In 2012 the European Organization for Nuclear Research (CERN) believed they found it. Marko Rodin, an independent researcher, claims he has it. To fully understand what they claim, one must first understand the basic principles of the theory of everything, also known as the grand unified theory, or holy grail of physics.

QUESTION: What exactly are being unified? What is "everything?"

ANSWER: Everything can be summed up in energy because we know Einstein's General Relativity, $E=mc^2$. Energy equals matter times the speed of light squared. In layman's terms, E= everything that moves. Through scientific research we know inanimate objects vibrate at the atomic level. Everything moves. Energy can be further broken down into four standard forces: gravity, electromagnetism, strong nuclear force, and weak nuclear force. To unify everything in the universe we must begin with these forces.

QUESTION: How and why should we unify everything into one equation?

ANSWER: Because doing so answers once and for all the universal question, who am I?

At some point in our lives we run into the "who am I's" and we begin to open up our mind to science, religion, and many other ideologies. "Where do we come from and where do we go once we die" are existential questions every human being will eventually ask after we're born and once we have self-awareness. Often children will ask these questions, and the parent will find themselves searching for the right words. Many of them will give a trite answer, suspecting their child will be satisfied. I think we underestimate the ability a child has in understanding the truth behind this answer. Many children grow up seeking answers by paving their own path in pursuit of a satisfactory explanation. We're drawn to the sophisticated idea of this complex puzzle. In some form or fashion, we associate ourselves with a point of view because without doing so we compromise our self-worth. I know I did. Religion and science are the two mainstream forces behind the influence of belief systems. The origin of species is the prize, and for centuries religion and science were at war over this. However, recent history reveals more and more people are open to the idea of joining these two into one belief system. It seems people are getting fed up with the complexities of this and that, and the do's and don'ts, and the "he-said she-said" explanations.

Everything has been over-complicated, and it is time to finally clean house. I'd like to sweep away all the dust and cobwebs so we can see everything for what it really is. In order to break it all down we will need to loosen the particles of dirt within the cosmos of our pre-conceptions with some heavy-duty cleaning spray, or better yet, *Comet* (hehe).

With a clean slate, let's define the four standard forces in no particular order.

1. Strong nuclear force
2. Weak nuclear force
3. Gravity
4. Electromagnetism (EM)

Quite simply, as you know, electromagnetism lights our houses. Gravity keeps us on the ground. Weak nuclear force is radiation... the process of decay and x-ray machines are good examples. Strong nuclear force is exhibited by nuclear bomb detonations and is also the same force holding everything together... our {subatomic} atoms together. Our sun is one huge thermo-nuclear reactor and serves as one example of strong nuclear force.

Approximately as little as a hundred years ago there were only two known forces: Gravity and electromagnetism. We hadn't yet discovered how to make nuclear weapons and we hadn't realized that the sun was a nuclear reactor either. Also, we hadn't discovered the radioactive properties of certain isotopic elements such as Uranium and Plutonium, so we were unacquainted with slow nuclear force even though it was happening naturally around us. Decomp. Decay was *not* unknown to us, but its radioactivity was. *Isotopes are variant elements of the same elements (more in a later chapter).

We were ignorant to the Strong Force until the nuclear bomb was created under the leadership of Robert Oppenheimer of the Manhattan Project during WWII. It was only after the fact, we learned, that scientists realized our sun and all suns and stars in the universe are natural sources of strong nuclear force akin to the force they bottled in atomic warheads. Strong nuclear force is the most destructive of the four forces, and that is why it is called S*trong*.

Weak Nuclear Force is radioactive decay. Essentially, it's the same thing as the Strong but it is released in smaller amounts. Take a balloon for example and fill it with toxic gas. If the balloon pops it released a *strong force* of toxic gas. But, if you hold the opening just so it only leaks a little bit, you have established a *weak force* of the same toxic gas. Same stuff, different delivery. The weak force can be found in compost and all dead matter in small doses. Everything that dies becomes toxic waste, and radioactive to some degree. Then, as aforementioned, there are the radioactive elements, such as isotopes of uranium and plutonium, a segue which brings us to the original discovery of the weak force. Scientists did not realize *Weak's* existence until they could measure it, and they didn't begin to measure it until a French-schooled Polish woman from Russia entered the scene.

At the turn of the 20th century Madame Marie Curie (with her hubby Pierre) discovered two new "radioactive" elements, Radium and Polonium. Polonium, she named after her homeland, Poland, a tribute to her nationality. Their research rewarded them as they both became recipients of Nobel Prizes. Madame Curie is the only woman to be a Nobel Laureate and the only person to have received prizes in two separate fields of science. She, also, coined the term *radioactivity*. Another radio-active element discovered much later was named after *her*, Curium. It was after their discoveries that the world of science was able to determine this third force as unique and powerful, however deceptive. Although Marie Curie's research gave science a quantum leap forward and earned her remarkable honor, there was a price to pay. She died of cancer. Continuous exposure to radioactive elements took a toll on her unsuspecting body. Slow nuclear force is extremely volatile even though you don't notice the effects. Like having a chest X-ray, you don't feel the effects of the force, but it's there and it's deadly if you are over-exposed to it. Radioactive fall-out from the atomic bombs dropped on Hiroshima and Nagasaki were devastating. Victims reported not knowing or feeling anything before the symptoms arose and created more casualties of war. The lethal effects of radioactive substances have a force to be reckoned with, even though it moves at a snails' pace. Hence the name Slow or Weak Nuclear Force.

But as it were, there were only two known forces before the 20th century:

1. Gravity
2. Electromagnetism

Prior to the 19th century there were three known forces. If you're paying attention you're probably asking yourself how's it possible to go from three to two forces... and then to four.

Before the humble Englishman, Michael Faraday, entered the picture in the late 18th century electricity and magnetism were known as two forces *separate* from each other, giving us:

1. Electricity
2. Magnetism
3. Gravity

Michael Faraday UNIFIED them. He unified the forces of electricity and magnetism when all the forces, all the odds were against him. He beat the odds and was the first to unify two forces into one. Michael was born into a world when philosophers and scientists were itching to unify electricity and magnetism. Prior to the 17th century and the laws given to us by Sir Isaac Newton, scientists were hoping to unify all three. Newton, however, shattered their hopes of ever doing so. But, it all began with a Greek mathematician, Thales of Miletus, born in 625 BC in Ionia which is located currently in the Turkish province of Aydin. Thales is one of the Seven Sages of Greece because he was one of the first philosophers following in the "Greek Tradition." This Ionian philosopher saw shards of iron sticking to lodestones. Later, the district of Magnesia was the first to mine lodestones, so they coined the term magnetism even though Thales made the discovery. By the 16th century an Englishman, William Gilbert found that fossilized tree sap known as amber attracted chaff and straw. This force was given the name Elektron (the Greek word for amber). That's how electricity and magnetism were first discovered. When I say discovered, I mean contemplated, because as you know it was Benjamin Franklin who first conducted electricity from lightning with his kite. But that was not until the mid-1700's when at about the same time electrical Leyden Jars were all the rage. Up until then scientists were clueless. Lodestones and amber. Hmmm. *Again, I must say it's fascinating that electricity was first realized this way while all along lightning was as evident as the sun. Although, I did see the other day a documentary of ancient artifacts. In it they revealed what appeared to be a battery and a light bulb. So, who knows! Perhaps electricity lit up ancient civilizations!

Gravity. Newton gave us its laws of motion, but it was Aristotle who coined the term. In his Magnum Opus, *Physics,* he referred to the *gravity of solid* objects as opposed to the *levity of gaseous* objects. And so it was before the 1800's the only three distinguished forces known to man for centuries, *millennia*, were:

1. Electricity
2. Magnetism
3. Gravity

It was obvious to scientists and philosophers that these three forces shared a commonality.

They were entertaining the idea of a trinity unified by one force, much like that of the Holy Trinity of the Christian Church. However, Newton came along and had apparently proved that gravity was unlike electricity and magnetism in that it only attracted objects while the other two forces were polar... they attracted *and* repelled.

The works of Nicolaus Copernicus and Johannes Kepler paved the way for Newton to deliver his laws of motion and crush their dreams of unifying the three forces into one "holy trinity." Ironically, Newton, being a Christian himself, unwittingly separated science from the church. By the 19th century, Charles Darwin settled the divorce between church and science which devastated the Christian paradigm. But at about the same time Michael Faraday unified electricity and magnetism and restored the hopes of fundamentalist philosophers in their belief of all forces having one origin. In modern science this belief or ideology is known as the Grand Unification, as mentioned above. The holy grail of science.

So it was that Michael Faraday unified electricity and magnetism, but it was James Clerk Maxwell who gave us the mathematic formula to Faraday's discovery. I find it such a shame that science gives Maxwell most of the credit, when it was Michael who really figured it out with simple copper wire and bar magnet. Maxwell has championed his own discoveries, such as light waves, which was a significant find and led to theories about the aether, which we will not get into here. I'm just say'n that Faraday having to share the limelight with Maxwell, or worse, be overshadowed, is disappointing. Michael deserves more credit. I'm a firm believer in giving credit where credit's due! At least Michael gets all the credit for his Faraday cage, for which most people recognize his name. Speaking of credit, my knowledge of Faraday and much information in this chapter comes from Michael Guillen, Ph.D., who first inspired me back

1997 to the poetry of mathematics, from his book Five Equations that Changed the World. I dedicate this chapter to him because I would not have been able to regurgitate much said here without him first telling it to me. I hold him in high esteem as we "share the same language" in passionate appreciation for mathematical linguistics.

The year was 1791 when Michael Faraday was born into a Sandemanian family outside London England. His father, an apprentice to a blacksmith, had a meager salary. Education was available only to those families who could afford it, and the Faraday's were not one of them. Where they were poor in finances, they were rich in spirit. Sandemanians were a devout people from a Christian sect in Scotland called the Glasites, after John Glas who founded it approximately 60 years before Michael was born.

By the time Michael reached his teens he had to find a job. That's how it was done back then; as soon as you were able, you were made willing to work to help provide for the family. You start in an apprenticeship, and lucky for Michael his was to a bookbinder. Now the hungry boy could get fat. The food was in the books, and the utensils were his eyes, heart and mind. His passion to learn compensated for his lack of education. Can you imagine… he must've devoured every book he could get his hands on… and he couldn't get enough of the latest scientific research. He must have soon realized he had a special interest in electricity and chemistry because before long he had a makeshift lab in the back of the bookbinder's shop where he conducted his own experiments. To boot, he made his own book beautifully bound with his own hands where he kept all his notes and findings.

The story goes, his book/journal was so beautifully made that one day (while Michael was not in the shop) an aristocrat came in to buy a book, but was lured by the beauty of Michael's journal sitting on the mantle, not for sale. As this gentleman thumbed through the pages of Michael's journal he became excited and insisted the bookbinder sell it to him. Captivated by the charm (and the wallet, I'm sure) of this distinguished man, the bookbinder sold it to him on the spot. Of course, I imagine when Michael returned to the shop to find his book gone he was totally devastated. I know how deflated he must have felt; this is akin to a computer crash! He lost it all. I've been there, lol. Worse for Michael, all his work was lost to him, but in the hands of a stranger.

Michael became so depressed that he nearly gave up on himself. He lost the motivation to read and to continue his own research. Then one day, out of the blue, the aristocrat came back to the shop with Michael's book tucked under his arm. He handed it back to Michael, so pleased to meet him, "the author," and with it the gentleman had an amazing gift for Michael. Bookmarked in the pages of his journal, for Michael, was a ticket to enter into the Royal Institute of Science to attend lectures of Humphry Davy. Davy was not only the chairman of the institution, but he was one of the most progressive scientists of the day in electro-magnetic research. Holy cow! Michael hit the jackpot! This was unheard of back in those days; a commoner such as Michael invited to attend one of the world's most elite institutions must've been similar to winning the megabucks lottery.

Humphry Davy was full of pomp and circumstance, and full of himself, to boot. Ever since he received the Bonaparte Prize from Napoleon via the Institut De France, his already over-inflated ego ballooned to zeppelin proportions! In spite of England's third coalition against France during the Napoleonic Wars, Englishman Humphry received this French award because he built the largest Voltaic Pile in 1807. Years earlier, the Voltaic Pile was first formed by Alessandro Volta from Milan Italy. Volta, essentially, built the first cell battery, hence the name. He alternated copper discs with zinc (serving as the electrodes) separated by Cardboard soaked in brine (salt water, which served as the electrolyte) and stacked them high one after the other. Alessandro Volta was outdone by Humphry when he made one so high and powerful that as a result sodium and potassium were discovered. Later, he found more elements using the Voltaic pile; barium, calcium, and magnesium, to name a few. He would hold scientific shows held in lecture auditoriums for exclusive groups of the rich and famous. Humphry regarded himself indistinguishable from royalty. And yes, he was a royal piece of work, and a royal pain in the ass, and it didn't help when the queen knighted him Sir Humphry Davy.

Sad to say but not surprisingly, Michael was not treated as an equal, not in the beginning anyway and it took many years for him to gain the recognition and respect he so deserved. After attending the free lectures by Humphry Davy, Michael sent to Davy 300 pages of notes that he took while he was there. Davy, in turn, gave Michael a job at the institution as his assistant. Later, on the road touring Europe, Michael was treated as a servant inside the lecture hall and outside of it as well. In the caravan, he traveled with the laborers and servants. In fact, Michael served two purposes: one as an assistant to Davy professionally, merely cleaning beakers, flasks, and petri dishes for Davy before and after his lectures, and the other as a servant to Davy as his personal valet (manservant). If it weren't for all the inspirational people he met during this tour in Europe and if it weren't for his amazing faith in God's provision, Michael would have probably called the whole thing off as his self-esteem was beaten to a pulp. Back in London at the Institute, Davy's treatment of Faraday got worse as the years went by. Michael's research and achievements were snubbed by a jealous Sir Humphry Davy, I'm sure, and encouraged by Humphry's wife, who was worse than he. The long and short of it is, although Michael endured hardships and was degraded to nothing more than a servant, his story has a storybook happy ending.

Michael ended up as the chairman of the Royal Institution! How awesome is that! He was esteemed one of the most influential scientists of his day while he was STILL alive. Go Michael! Also, during his golden years, he received a royal invitation by Queen Victoria to be knighted and have a place of burial right next to Sir Isaac Newton in Westminster

Abbey! Michael, the humble man that he was, respectfully refused both invitations. He declined the knighthood and simply wanted to be buried next to his father in his hometown, and so was he.

Faraday accomplished much for the world of science. For the many things he did, such as take a first look at nanoparticles, and the laws governing electrolysis, his most important and world-changing discovery is when he wrapped copper wire around a magnet. This seemingly super-simple experiment led to the unification of electricity and magnetism, which is why I am telling this story. Michael found that a current of electricity flowed as long as there was an increase or decrease of magnetic force. The secret was found in the "on and off" connection. Picture a rotation of a magnet. The faster the spin the stronger the current, the slower the magnet rotated the weaker the current. And if there was no change at all in the magnetic field, nothing happened. As a result of Faraday's discovery dynamos were created, the very first electrical generator. Dynamos eventually replaced steam engines and were the powerhouse of communities everywhere. Later, bigger and faster rotating magnets within Dynamos were the electric source of Thomas Edison's bulb and Bell's telephone. It was Nikola Tesla who, less than a century later, discovered the rotating magnetic field which gave us the AC (alternating) current which was used by Westinghouse to replace Edison's and Alexander Graham Bell's DC (direct) current. These amazing discoveries made it possible for all of us today to have electricity in our houses and workplace.

Officially, Faraday's discovery back then changed the model of standard forces from:

1. Electricity
2. Magnetism
3. Gravity

TO

1. Electromagnetism
2. Gravity

You see! Electricity and Magnetism were unified into one of the now four standard forces.

Michael did not know how to put his findings into mathematical language, so he wrote the following: "Whenever the magnetic force increases or decreases, it produces electricity; the faster it increases or decreases, the more electricity it produces."

A short time later Michael's discovery was put into a formula by James Clerk Maxwell.

$\nabla \times E = -\partial B/\partial t$

*The inverted triangle ∇ represents the "amount of" and the E represents "electricity." B represents "magnetism" and $-\partial B/\partial t$ stands for the "rate of increase or decrease of."

So, there you have it. Now you know how Electricity and Magnetism (originally known as two separate forces of nature) were unified into one.

Let's recap:

Prior to Michael Faraday's Discovery there were THREE STANDARD FORCES:

1. Electricity
2. Magnetism
3. Gravity

After Faraday's discovery and Maxwell's equation there were TWO STANDARD FORCES:

1. Electromagnetism
2. Gravity

Today there are FOUR STANDARD FORCES thanks to Madame Curie and Pierre, and (no thanks) to all atomic bomb engineers:

1. Gravity
2. Electromagnetism
3. Weak Nuclear Force
4. Strong Nuclear Force

From three to two to four! However, we are becoming ever so close to: From Three to Two to Four To ONE!

*To learn all the details of Michael Faraday's life, please read Five Equations That Changed the World. Dr. Guillen tells it best! He is one of my favorite storytellers and an inspiration to me.

Dr. Einstein spent the last 20 years of his life trying to unify the current four standard forces into one. He had especially concentrated on gravity. In his "latter" years, Einstein said, "I am now working exclusively on the gravitation problem... but one thing is certain: never before in my life have I troubled myself over anything so much... compared with this problem, the original theory of relativity is child's play."[1] Einstein died trying, however, he was able to find unity between strong and weak nuclear forces, but today they remain as two separate forces.

The four fundamental forces have different dynamics in their functions but have long believed to come from one origin. This belief is the motivator for scientific minds to unify them mathematically.

Have we found it? CERN thinks so with their Higgs Boson, a subatomic particle they discovered using their atom smasher, the Large Hadron Collider (LHD (more on LHD in the next chapter and more on Bosons in chapter 11)). Marko Rodin, according to his self-proclaimed protégé, Randy Powell, believes he discovered it when he stumbled upon a unique number pattern in the torus. He called it vortex mathematics and dubbed his finding as the Rodin Coil.

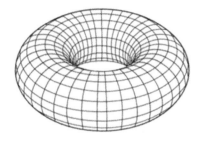

Figure 1. Torus

In geometry, a torus is a surface of revolution generated by revolving a circle in three-dimensional space about an axis that is coplanar with the circle. If the axis of revolution does not touch the circle, the surface has a ring shape and is called a torus of revolution.   - https://en.wikipedia.org/wiki/Torus

For more information check out: https://www.chegg.com/homework-help/questions-and-answers/torus-also-known-donut-mathematical-object-formed-revolving-region-bounded-circle-x-2-y-2--q21852685

In the chapters ahead we will examine these findings and their claims as well as take a trip down memory lane. The lane I treaded was full of pain-inducing obstacles which caused me to sink and nearly drown in dread and then teleported me to quantum constructs and from there I was risen up to sit aloft angelic clouds. This journey began as a chase which slowed to a stroll and then to a complete stop. Now it is a race to the finish!

The Limited

What is the standard
of your endeavor?

Whence doth it meet
its threshold?

2000

Chapter THRΣΣ

The Process of Illumination

Many anti-theistic scientists are eager to find the grand unified equation in hopes to disprove the creation theory. And many scientific theologians are split in this endeavor. Some are ambivalent because they fear they will have to battle the scientific findings in order to preserve their faith. While others adamantly believe no such proof even exists and, therefore, they have nothing to fear. Still some others believe God is the ultimate scientist so it would be impossible for any equation, if found, to contradict creation. They feel solid because there are so many theories and models to support creation. Cosmology, for one, which is my favorite because it is the most obvious. The simplest answer to anything is usually the correct one. Isn't that the simple truth of Occam's Razor? What's Occam's Razor, you ask? We will learn more about it later, but if you can't wait, *google it,* if you must. The Cosmological Model states that our universe and our solar system mimics a timepiece, and therefore, there must be a watchmaker. Do the teachings of evolution undermine all models of creation?

Pre-modern-day Madness

After more than a decade since I expeditiously followed the latest findings of particle physicists, I had read in current Science magazines (at that time) that they hadn't had much progress. These physicists, in their efforts to smash sub-atomic particles, including their best joint effort (to capture, measure, and study dark matter) were crashing their own heads together! I couldn't help but chuckle to myself because these guys were/are so incredibly smart. Their intelligence is astounding, and they seem to have an endless supply of sponsorship funds for their experiments. Consider the atomic accelerators they built… these machines are impressive as they are able to accelerate particles to near light speed to smash them together and break them apart, so they can take pictures of the smaller particles inside. These people are highly respected scientists, some holding multiple doctorates, and they get to play with these massive machines every day! *I can't help but feel a little jealous, like a kid out of the sandbox.* The Large Hadron Collider (LHC) in Geneva, Switzerland, at CERN, is 27 kilometers in circumference and 100 meters underground. That's more than 16 ½ miles around, and more than 109 yards below! Wow! CERN has several accelerators, but the LHC takes the record for energy levels… it is the mega-machine of the world having collided heavy ions of lead! See home.cern for more information.

*Fun Fact: CERN created the world wide web in 1989

Now imagine this: In Waxahachie, Texas, approximately 30 years ago, excited physicists were raising funds to finish building the Superconducting Super-Collider which would have spanned more than 87 kilometers (more than 54 miles round) and would have been the world's strongest/largest accelerator. In this case their sponsor's pockets weren't deep enough and congress cut their support funds, so construction stopped in 1993. CERN, the European Organization for Nuclear Research, remains the "big kahuna" and gets all the bragging rights for having the world's largest accelerator. More so, it's the world's largest machine. Although, fyi, the Relativistic Heavy Ion Collider (RHIC) in Upton, NY is a heavy contender! *Hehe, fun pun. We, Americans, have accomplished quite a bit with our own smaller accelerators. At Stanford, we found light produces matter! Holy smoke! No joke. More later on that... that "matter" warrants its own chapter. ☺

A physicist named Dr. Leon M. Lederman (who passed away in 2018, RIP), was one of my personal heroes. Back in the nineties I read his book called *The God Particle*. Although many of you may not have heard of him, he is a respected experimental particle physicist, who along with Melvin Schwartz and Jack Steinberger won the Nobel Prize in Physics in 1988 for discovering the muon neutrino, among other things. He was the Director Emeritus of Fermi National Accelerator Laboratory (Fermilab) in Illinois. Dr. Lederman was an integral part of the Physics First Movement, which I wholeheartedly believe in... because I, for one, would not have come to my own understanding of math without first having the revelation I did in physics. Allow Wikipedia to explain *Physics First*:

> **Physics First** is an educational program that teaches a basic physics course in the ninth grade (usually 15-year-olds), rather than the biology course which is more standard in public schools. This course relies on the limited math skills that the students have from pre-algebra and algebra I. With these skills students study a broad subset of the introductory physics canon with an emphasis on topics which can be experienced kinesthetically or without deep mathematical reasoning.
>
> Physics First began as an organized movement among educators around 1990 and has been slowly catching on throughout the United States. The most prominent

movement championing Physics First is <u>Leon Lederman</u>'s ARISE (American Renaissance in Science Education).

Many proponents of Physics First argue that turning this order around lays the foundations for better understanding of chemistry, which in turn will lead to more comprehension of biology. Due to the tangible nature of most introductory physics experiments, Physics First also lends itself well to an introduction to inquiry-based science education, where students are encouraged to probe the workings of the world in which they live.

The majority of high schools which have implemented "physics first" do so by way of offering two separate classes, at two separate levels: simple physics concepts in 9th grade, followed by more advanced physics courses in 11th or 12th grade. In schools with this curriculum, nearly all 9th grade students take a "Physical Science", or "Introduction to Physics Concepts" course. These courses focus on concepts that can be studied with skills from pre-algebra and algebra I. With these ideas in place, students then can be exposed to ideas with more physics related content in chemistry, and other science electives. After this, students are then encouraged to take an 11th or 12th grade course in Physics, which does use more advanced math, including vectors, geometry, and more involved algebra.
<u>https://en.wikipedia.org/wiki/Physics_First</u>

I wanted to share with you the importance of physics first not because it pertains to my book, but simply because I agree with its importance to better understanding science in general. It was a brief digression, but worth it. *Fyi, I will make tangents along this journey when I feel led to share the importance and not necessarily the significance of the subject matter. Or simply because I want to jump in the *rabbit hole* for giggles.

Back in the early nineties, Dr. Lederman was on a committee to raise awareness of and funds for the Superconducting Super-Collider (SCSC) so they could, in essence, do one thing. Find a grand unified equation. To find this equation was to find the holy grail of physics, he said. Michio Kaku, said the same thing, if my memory serves me right. His was the other book I read at the time, *Hyperspace*. Both Dr. Kaku and Dr. Lederman have written many more books/publications since then, but for the two of them, their first books, *The God Particle*, and *Hyperspace* were the most impactful to me in understanding this subject.

The holy grail of physics was the cup all physicists wanted to drink from... the contents, they hoped, would give them a picture of life. They didn't believe it would give them eternal life, but they were confident they would get a glimpse at the origin of life. While aging elites were searching for a fountain of youth, a scientific rat race to find a special particle was underway. They were all digging... some digging deeper than others. Their "digging" into particles with big machines had a sole purpose. For Leon Lederman, he and his colleagues were sure they would find it with the SCSC. At the time everyone seemed to believe they would find it in the Higgs Boson which was located in the Higgs Field.

The Higgs Boson is a sub-atomic particle, theoretical at the time in a theoretical field. In The God Particle, Dr. Lederman wrote that the sole purpose of building the estimated eight-billion-dollar machine was to find the Higgs Boson. They were confident that the collision of the correct composite particles the Higgs would show himself in collateral damage. It would be found in the rubble of the exploding elementary particles. Sounds like searching for a needle in a haystack, doesn't it? Today, CERN writes about it and tells of its nature in its Higgs field. The LHC succeeded in excavating it when the SCSC failed to even try. I imagine CERN's discovery was bittersweet for Dr. Lederman; I felt his eagerness to find it himself while reading his book. I'm sure he was happy to hear of its discovery, yet bummed it wasn't him who discovered it. I haven't looked into Fermilab's role, if any, in its discovery, but I did learn they do have an ongoing camaraderie with other atomic-research laboratories. They all share notes, it appears, so I wouldn't doubt if Dr. Lederman had some hand in, before or after, its detection by CERN in 2012, the same year he retired from Fermilab.

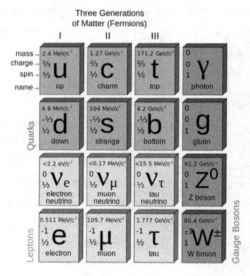

Figure 2. Standard model of indivisible particles (older version)

Looking at this graph of the atomic model (see figure 2) of particles you see sections which divide the particles into groups. The quarks and leptons have pairs, and the leptons are further divided into generational subgroups. To the right you will find the particles which carry the forces we learned about in the previous chapter. All except gravity. Photons and electrons carry the electromagnetic force and the gluons which hold all the particles together carry the strong force. Notice at the bottom right are two squares, W and Z. These Gauge Bosons carry the weak force. We will return to this model and learn more about these particles in chapter 11. For now, let's just take advantage of this chart as a model to understand the building blocks of atoms which are the building blocks of molecules, so we can stack them all up, like Jenga, and knock them over!

In the 90's after I first learned of their (scientists) intentions to find the holy grail via mega-machines, I was amused by their exorbitant and expensive ambition which I was sure would wind up fruitless. "I can't imagine that their collective data has accelerated their results only a few microns!" I'd exclaim derisively as if I were qualified to do so. I became paranoid others would think I was a megalomaniac, so I began to keep these thoughts to myself. I'd think quietly and tell myself, "combining their genius, from engineering feats to theoretical research, you would think we would have a clear picture of dark matter by now. You know, the perfect shot of *nothing*. At least they've confirmed that neutrinos move faster than the speed of light." I'd say these things because I felt so strongly that they were wasting time and money to find the Higgs Boson. I tried to reach out, but no one cared to hear what I had to say about it. Who can blame them? My attempt was as futile as theirs. I was a nobody in science. Still am. Since then, CERN has made gigantic strides, touting the Higgs Boson as the God Particle, and I humbly admit they have indeed found something out of this world (more on this later).

Neutrinos are relatively the smallest particles we've found. How small is a neutrino? Neutrinos are roughly a millionth of the mass of an electron. How big is an electron? Picture a regular atom; a hydrogen atom, for instance. An electron is a subatomic particle orbiting around the nucleus of the atom. I'm assuming you already know how small an atom is, generally speaking... they all come in different sizes, or atomic weight, but for all intents and purposes, use your imagination to close in on an atom, any atom, regardless of its mass. Just for those of you who do not know how small atoms are, they are the elementary building blocks that make up the molecular structure of all things. Waaay smaller than the cells Medical Examiners see when looking through a microscope. Waaaay. According to Wikipedia, "The upper molecular weight limit for a small molecule is approximately 900 Daltons which allows for the possibility to rapidly diffuse across cell membranes so that they can reach intracellular sites of action." https://en.wikipedia.org/wiki/Small_molecule

Daltons are units of measurements used to determine atomic mass. A Dalton is more specifically known as AMU, the unified Atomic Mass Unit.

Now before we go any further, allow me to explain the origin of the atom as we know it.

Democritus, an ancient Greek philosopher (460-360 BC), is credited for defining the ATOM. Leucippus, his teacher, shared many same ideas regarding science, so who's to say the credit should not begin with him. In short, Democritus described the atom to be "that particle which cannot be divided." He said if you cut in half a piece of something and then cut that piece in half, then cut that piece in half, so on so forth, you would reach the atom once the half is no longer divisible. He was philosophizing on a molecular level, so with his imagination, he would envision a microscopic knife with which he could continue to cut pieces in half. Not surprisingly, he did not come up with an exact measurement of how small one could cut, for the final cut. Even with nanometers (nm), angstroms (Å), and picometers (pm) used today, it would be impossible to hone in on what, when, and where we would stop cutting, right? Yet, the atomic model above supposedly contains some indivisible particles.

Molecules are the smallest unit of any chemical compound and are made up of atoms (see figure 3 for a variety of common chemical substances and its molecular structure). Notice, baking soda is made up of the chemical compound of a sodium (Na) atom + bicarbonate (HCO_3) which is made up of one carbon, one hydrogen and three oxygen atoms. The atoms are the smallest unit of elements found in the Periodic Table of Elements. The subatomic particles make up the atom, and atom smashers are digging into those tiny particles to look deeper.

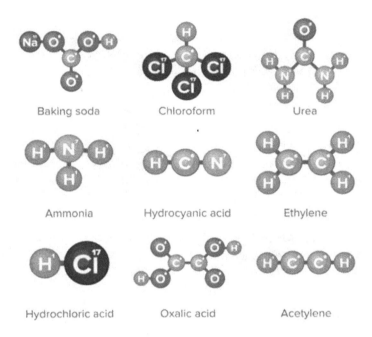

Figure 3. Molecules

According to Dictionary found on Google, "Molecules are a group of atoms bonded together, representing the smallest fundamental unit of a chemical compound that can take part in a chemical reaction."

In figure 4 we see atoms (which make up molecules)… they are held together by chemical bonds by sharing and/or exchanging their electrons. The bonds they make with other atoms are chemical compounds as shown in figure 3. Electrons are governed by the Electromagnetic Force, and they orbit the nucleus, made up of protons and neutrons, which are made up of smaller subatomic particles as we saw in figure 2.

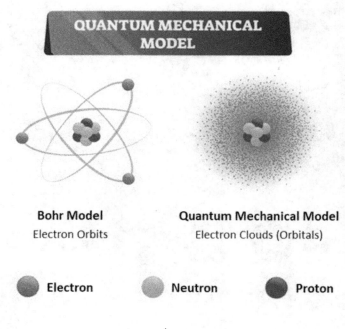

Figure 4. Atom

What would a unification equation look like?

In his book, The God Particle, Dr. Lederman surmised that the Grand Unified Equation would be so small that you could fit it on a t-shirt. As aforementioned, The Grand Unified Theory, or Holy Grail will give us the theory of everything. Back in the 80's and 90's when the race for "the grail" was in full steam, a British physicist who was also in the race, had his students join the marathon.

Professor Stephen Hawking said in his book, A Brief History of Time, that he received so many formulas from students that his mailbox could not contain them. *Back then it was a real mailbox, not email! Lol!* Hawking said that he could tell at first glance if a student was onto something by the size of the formula. One after the other, he would discard all the proposed formulas because they were too long. The formula would have to be small, like E=MC^2. Albert Einstein's equation of General Relativity (GR) is a wonderful formula unifying matter and light with energy. Einstein's field equation can fit on a t-shirt too, but it is not as small as GR. Then there's Maxwell's equation from Faraday's discovery, shown earlier; another t-shirt equation. There are numerous t-shirt equations out there. Here are a few:

Rudolph Clausius' equation on the 2nd law of thermodynamics:

$\Delta S_{universe} > 0$

Ohm's Law: $V = IR$

Heisenberg's Uncertainty Principle: $\Delta x \Delta p \geq \dfrac{\hbar}{2}$

Euler's Formula: $e^{i\pi} + 1 = 0$

Newton's Gravity: $F = G \times M \times m \div d^2$

And Newton's 2nd Law is even smaller: $\vec{F} = m\vec{a}$

Wouldn't it be fun to put Pi on a t-shirt? I can see it now; it would go round and round and round the shirt...
π=3.14159265358979323846264338327950288419716939937510582097
4944592307816406286208998628034825342117067982148086513282 30
6647093844609550582231725359408128481117450284102701938521 10

5559644622948954930381964428810975665933446128475648233… and on and on…

The T-Shirt Equation of the grand unification formula, as it was fittingly called, became THE Easter Egg every scientist and aspiring scientist was looking for! The world of science has seen so many theories and equations, but now the race seems to be over because CERN found the Higgs. The marathon still has participants walking its course because they still cannot formulate an acceptable equation. You see, the equation has to match the explanation. As we saw earlier, Faraday's original formula was not written with numbers and symbols because he did not have the learned math language for it. Maxwell simply interpreted Faraday's explanation into another language. Arithmetic. In order to achieve the desired T-shirt formula, one must have an explanation simple enough for the translation. So, when Stephen Hawking would see a formula someone would mail to him he could tell immediately that the explanation was extraneous; therefore, it couldn't remotely be *THE* equation. The equation has to paint a simple picture and be easy to understand. A grand unified equation demands a grand simplified explanation. Even for a fourth grader.

During the process of unification, things get smaller, not larger. The definition becomes compressed or reduced rather than expanded. You take the complexities and unravel them until you are left with the fundamental parts that pieced them together. A perfect example of this is what Jesus Christ did with the 10 Commandments. He unified them. Jesus took the Old Mosaic Covenant and simplified it into the New Testament Covenant. He reduced 10 laws into two!

You will find recorded in the book of Matthew 22:34-40:

"Hearing that Jesus had silenced the Sadducees, the Pharisees got together. One of them, an expert in the law, tested him with this question: "Teacher, which is the greatest commandment in the Law?"

Jesus replied: "'Love the Lord your God with all your heart and with all your soul and with all your mind.' This is the first and greatest commandment. And the second is like it: 'Love your neighbor as yourself.' All the Law and the Prophets hang on these two commandments.'"

The original 10 Commandments can be divided into two categories, so for convenience and functionality let's look at them the way Moses would have presented them... on two stone tablets. This may be a good time to look up the original 10 commandments in the Bible if you do not know them already because I will not list them here. You can find them in Exodus 20.

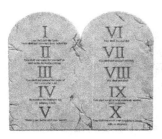

Figure 5.

Jesus said that if you truly love God you will automatically obey everything on the left side of the tablets (the God Group). Likewise, if you truly love people you will obey everything on the right side of the tablets (the People Group). *Some argue that the 5th commandment should be categorized with the right side; that honoring your parents falls in the people group and not the God group. I believe God purposefully chose to include the 5th commandment with the left tablet because parents are a metaphor of God the Father. He uses a parent's love over and over again in the Bible to describe His own love for us. God also uses our birth lineage as an example of the inheritance we will receive through Him (Yeshua).

The birthright and connection we have to God is in our human DNA tracing all the way back to Adam and Eve. God says we are made in His likeness and in His image. But that's beside the point. Bear with me, I will bring this all together in the end. I'm painting a picture, and these are the first strokes. A later chapter will reveal the image.

Notice that Jesus, when questioned by the priest, not only summed up the Mosaic Law (The 10 Commandments) but he included all the laws of the prophets, known in Hebrew as the TaNaKh. Really, Jesus unified more than ten items because there are 613 Old Testament commands or laws. This is an example of unification. If I were to translate it into a symbolic language it would look something like this:

♥AΏ x ♥P = ΣOTL

This can be translated as *Love of the Alpha and Omega times love for People equals fulfillment of Old Testament Laws*. I played around with this for a while and had fun with all the different symbols and combinations. I think this is the simplest... from a human "girly" standpoint. Lol! You get the picture!

From God's standpoint this can be broken down even further. Jesus said first we must Love God with everything that is in us. If we do that we can establish a foundational relationship vertically toward God. Heavenward, ↑. In addition, if we love the people around us we can establish a relationship horizontally with those around us. Side by side. Hand in hand. Outward, ↔. Once you combine the two you have the cross, † which, as you can see, is an even more unified expression of the Old Testament laws! The cross is significant to many things. It's not just a religious symbol of Christianity.

Is it any coincidence that Laminin looks like a cross? Laminin is a glycoprotein that holds everything together in our bodies. They are cell adhesion molecules. Webster's Medical Dictionary refers to Laminin as a component of connective tissue basement membranes. Louis Giglio said it best when he described Laminin as the rebar of our lives. Before you pour concrete into planned foundation of any new construction you must first insert the iron rebar in order for the concrete to have something to bite into, he said. Laminin, like rebar, keeps the foundation of our cells from falling apart. Without it the matter that makes us who we are would collapse before our very eyes. Laminin has been dubbed as the God Molecule, for obvious reasons. Not only is it the cement for the foundation of our lives, but the molecule itself is in the shape of a cross. Figure 6 is a copy of a photo made famous by Louis Giglio.

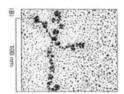

Figure 6. Electron microscope image of mammalian
laminin in a cross formation

There are other photos of it where it does not appear to be a cross... thereby naysayers have claimed it to be a hoax or have criticized Louis for calling Laminin a cross. Here is my rendering of Laminin revealed in figure 7 as a "non-cross." To see an original microscopic image, just Google it. Plenty of images pop up as a result.

Figure 7. Mammalian laminin in a non-cross formation

Laminin, unlike rebar, is living and malleable, therefore, every picture of it will *not* be the same. It moves as the cells move. I don't think Giglio made any proclamations to say otherwise. He did not claim Laminin is always [viewed as] a cross, as some have accused him of and criticized him under those assumptions. Whether a laminin molecule is captured by photograph posing as a cross or a hook or a figure eight or whatever, is beside the point. The schematic structure of the molecule is a cross, nonetheless... see figure 8, lol!

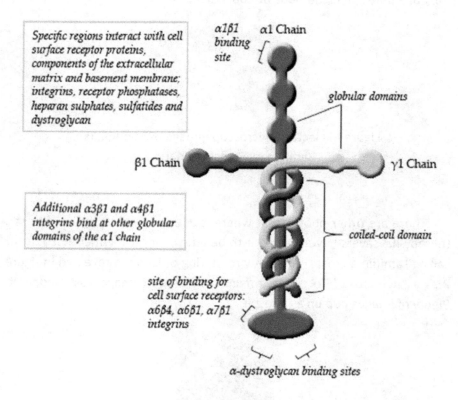

Specific regions interact with cell surface receptor proteins, components of the extracellular matrix and basement membrane; integrins, receptor phosphatases, heparan sulphates, sulfatides and dystroglycan

α1β1 binding site

α1 Chain

globular domains

β1 Chain

γ1 Chain

Additional α3β1 and α4β1 integrins bind at other globular domains of the α1 chain

coiled-coil domain

site of binding for cell surface receptors: α6β4, α6β1, α7β1 integrins

α-dystroglycan binding sites

Figure 8.

Even so, naysayers (atheists) continue to deny the cross. They prefer to view it as a caduceus:

Figure 9. Caduceus

Is it a coincidence that those who oppose the cross, oppose Christ? And those who oppose Christ are called antichrists. And is it any coincidence that those who oppose Christ *prefer* a symbol with a snake? Two of them! Btw, Snopes says "false" are the claims of Laminin as evidence of God. *Thanks Snopes! You would know! So glad you cleared that up for all of us. What would we do without you!*

The structure of Laminin can also be viewed as a sword when turned 90° as in this photo:

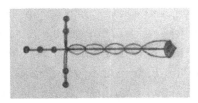

Figure 10. Laminin molecule crude diagram

Figure 11. Medieval sword

Further still, is it any coincidence that Laminin may be viewed as a cross or a sword, and that they are both held in spiritual regard by New Testament authors? Just as the cross is more than a prop for Roman execution, the sword is more than a weapon to, likewise, kill. In Ephesians 6:17 the Apostle Paul tells us the sword of the spirit is the word of God and in Hebrews 4:12 we are told that the word of God is sharper than a two-edged sword, able to cut between joint and marrow, and between soul and spirit. In John's first chapter we are told that, "In the beginning was the Word and the Word was with God. And the Word was God..."

Is it possible that the sword within our cells can cut the "rope" that holds us together, and perhaps before the fall of Adam and Eve in the Garden they were able to move between physical and spiritual dimensions with ease? Walk through walls, so to speak... or in their case, walk through trees. Lol! Is it possible that after they bit into the apple of death they were no longer capable of this human function? As I recall the account in Genesis 3:7 as soon as they ate of the fruit it says their eyes were opened and they sewed fig leaves together to cover their nakedness. Then in Genesis 3:21 it says that God Himself made them leather clothing after Adam admitted hiding from God due to his nakedness. How did Adam and Eve know they were naked? God asked them the same question. All Adam could do was blame Eve, and all Eve could do was blame the serpent. Not much else is said in these verses, but enough was written to determine that before they "ate the apple" they were not naked, but afterward they found they were! I do not believe they were, all of a sudden, ashamed of their private parts. I believe they must've had a covering. Clothes spun with threads of light! Is it possible they were bodies of light... controlled somehow, in part, by Laminin? Let's take a closer look at the verse, aforementioned, in Hebrews 4:12...

^{12} For the word of God *is* living and powerful, and sharper than any two-edged sword, piercing even to the division of soul and spirit, and of joints and marrow, and is a discerner of the thoughts and intents of the heart. [13] And there is no creature hidden from His sight, but all things *are* naked and open to the eyes of Him to whom we *must give* account.

Yet, another coincidence is found here in the subsequent verse of the double-edged sword where you find the Apostle Paul saying that when we come face to face with God we will be naked. Made totally bare before the One who will judge us on judgment day? Just the mere possibility of that makes me shiver. Is this only a spiritual metaphor? I wonder if Paul is referring to nakedness both inside and out? Totally exposed body and soul? This verse brings to life, at least for me, the verse from Proverbs 9:10, "the fear of the Lord is the beginning of wisdom…" If God has the power to strip us naked we'd be wise to hang on His every word lest we find ourselves in an embarrassing situation, much like Adam and Eve. Just say'n.

Let's take a look at 1 Corinthians 15, starting at verse 50…

[50] I declare to you, brothers and sisters, that flesh and blood cannot inherit the kingdom of God, nor does the perishable inherit the imperishable. [51] Listen, I tell you a mystery: We will not all sleep, but we will all be changed— [52] in a flash, in the twinkling of an eye, at the last trumpet. For the trumpet will sound, the dead will be raised imperishable, and we will be changed. [53] For the perishable must clothe itself with the imperishable, and the mortal with immortality. [54] When the perishable has been clothed with the imperishable, and the mortal with immortality, then the saying that is written will come true: "Death has been swallowed up in victory."[a]

[55] "Where, O death, is your victory? Where, O death, is your sting?"[b]

[56] The sting of death is sin, and the power of sin is the law. [57] But thanks be to God! He gives us the victory through our Lord Jesus Christ.

Is it not entirely possible that those who follow Christ, who will be "raised up" with Him will have on garments of light? It says they will be "clothed with the imperishable" in the "twinkling of an eye." What if, in an instant, their Laminin as well as other circuits in the body will be turned on, as if God flipped the switch!? Joint and marrow will be divided. Soul and spirit will be cut, and light will shine though allowing them to move from the physical into the spiritual. In Revelation 20 it says blessed are those who are part of the first resurrection, as described above and, also, in 1 Thessalonians 4:13-18, this event is the first resurrection. It says that death has lost its sting and the 2nd death has no power over it. Remember, above in Hebrew 4:13, it says we will all be naked before HIM. Everyone except those in white garments according to Revelation! "The bride of Christ," clothed with the imperishable as it transformed from earthly bodies to spiritual bodies in a twinkling of an eye! Two groups. Clothed vs naked?

Laminin is holding our cells together. It's more than just cellular glue. It is the rebar in the foundation of our flesh. Basement membranes, it's called. What if it becomes supernaturally active, sharper than a two-edged sword, and severs the connection between our cells, will the crude molecular matter of our physical existence crumble, and we would become as light? We could walk through trees?

Regardless of what we believe and how we perceive it, Laminin's likeness to the cross and sword is uncanny. Snopes claims Laminin as evidence of God is *false*, but If we turn to Snopes for spiritual answers we are definitely and desperately in need of prayer, right? Truly, what are the qualifications and who or what is qualified to tell us the answers to the origin of life? There are two kinds of people in the world: one who believes a coincidence deserves further introspection, and the other who does not. If you tend to lean toward the former, you're a lot like me. If not, I trust by the end of this book you will begin to question, just like I do, why so many things in life are synchronistic and why the cross always appears at every *cross-section* of our choices. ☺

So far as I can tell from what we just dissected, a unified equation will not put a wedge between God and science. Also, the images of the cross are, coincidentally, similar to cell adhesion molecules and, ironically, to the cutting sword which give the cross deeper meaning not contradicted by Biblical teaching or science. I can appreciate this metaphysical approach to better understand my spiritual condition. This is fun! Let's keep going. Jesus unified the ten commandments into the new testament covenant, love God and love your fellow man. Some people are uncomfortable by commands, but this one given by Jesus (two-fold commandment) seems to be benevolent. God and man... appear as though the two are married into the ultimate T-shirt equation. But what does this have to do with science? And what and where is the key to unlock our molecular make-up? There's more. The cross represents a unified formula for mankind's relationship toward God and each other, and what we saw with the cross and sword has profound physical/spiritual implications. What we saw with Laminin was corporeal on a cellular level (attached to our corruptible flesh) but may have metaphysical functions (attached to our incorruptible spirit), as well. This can be broken down even further. We can keep slicing. We need to go atomic!

Jesus said, "...Blessed is the one who stays awake and remains clothed, so as not to go naked and be shamefully exposed." Revelation 16:15

The Road to My Soul

On-ramps are everywhere
Speed limit is one hundred eighty-six thousand
four hundred sixty miles per second
Mobile phones are prohibited
Sunglasses mandatory
2000

Chapter FΦUR

Runaway

"**I**f you do not come home immediately, do not come home at all," were the stern words my dad gave me when I was on the phone with him at my friend Avery's house. I had been there a week without my parent's approval, much less without them even knowing where I was. Come to think of it I cannot recall how my dad tracked me down in the first place... I would have stayed there indefinitely!

A week prior I was on another phone call, with my friend, Carrie.

"Can you come over?' She asked.

"No, my mom is being a b!#@h," I replied in a whisper. The hushed tone didn't help.

"I am going to kill youooooo...." came an incensed growl as it trailed off into what sounded like a fiery furnace. My coronary artery almost burst, and before I could hang up the phone, Oma was already in my room swinging punches at my face! She tore down all my drawings pinned on my wall like a caged tiger, roaring. Screeching. Ripping. Punching. Repeat.

I was outa there! I don't recall the details of how Oma came to a stop if she did at all. All I remember was jumping into Avery's Karmann Ghia and driving away in tears with years of brewing hatred in my heart coming to a rolling boil.

The week went by so fast, yet, at the time we did so much. At the time, fun stuff. In hindsight, sin stuff. We drove around, car full, squeezed like a pack of sardines jamming to Aerosmith or The Smiths if we weren't playing around on the CB radio. "Break one-nine for a radio check. Anyone out there want to buy us some beer?" There must've been six or seven of us in Avery's car. If we couldn't find someone to buy us spirits we stole it. 3.2 beer and wine coolers from the grocery store, stole liquor from the bowling alley, and while we were at it we stole clothes from the mall. Broke into Mile High Stadium to see a U2 concert. Drove around neighborhoods to see what car we could break into to steal stereos. We even went to a Mexican restaurant and stole sombreros from their reception area. I recall running out of there with our hand atop our humungous hats so the wind wouldn't blow them off. Piling back into the tiny Karmann Ghia, barely able to see, or even shut the doors for the hats were so huge, as we sped away!

At the time, I say fun stuff, but in my heart I knew it was wrong. In fact, several times I questioned myself, *why am I (why are we) hanging out with Max?* Max was the guy sitting shotgun, Avery's right-hand man. And it appeared to me that all the crime we were committing was led by his hand, and none of us had the presence of mind to rebuke all the shenanigans!

Sitting next to me in the back seat was Autumn, a church-going girl like myself. You would expect either she or I would have said something.

But, no. Not a peep. I know she must've been thinking the same thing I was, *are we really doing all this!* We were just as guilty. Not just by association, we were participating. I looked back on it with shame and guilt until the day I repented and took it to the Lord in prayer. I took a lot more than that to the foot of the cross. Much more.

By association and participation, I started smoking pot and experimenting with psychedelics, I lost my virginity and had not one but two abortions. I was a total reprobate. I walked the wide road of promiscuity, looking for love at every intersection, partying with a party of party animals. Between crossroads, I'd find a food stand with the devil handing out shiny red apples. I ate nearly every one. To say the least, I did not live a "wholesome" life. Surprise surprise! In fact, I stopped going to church…. my home church. I knew right from wrong, and I won't make any excuses save for the fact that my home-life was not very loving. In retrospect, my true experiences regarding love came from my grandmother Pearl, and the pastor's family of my home church. Sure, I mentioned how my dad would hug and kiss and tuck us in at night, with his "maternal" affection … I felt the love, but, by high school it must've been cancelled out by his physical absence and eventually by his emotional absence. As for Oma…, well what can I say. If she had shown any love, such as a hug or a kiss, maybe that would have balanced out the beatings, for my own memory's sake. The lack of love at home was compounded by my two brother's violent nature. They were the tornados of the storm where Oma was the hail! I always felt like I had to take cover. I recall once, my brother picked me up and threw me across the room, clearing my parent's king-sized bed and hitting the far nightstand painfully with my shoulder.

I was no saint, but I sure tried to act like one when I would manage to make it to church. Church. *What the heck was church, anyway! And why did I even bother to go?*

My earliest memory of church was in San Diego, CA., when I was four or five years old. I recall spending more time at the pastor's house, with his wife and two sons, then actually being in a chapel. I do remember Sunday School and going to a building with a chapel, but for the most part, the people in small gatherings make up most of my earliest recollections of church.

We moved to California before Zeke was born, three years my younger. My dad was stationed there from South Korea, where I was born. Oma worked at a bar in Itaewon, South Korea, a suburb of Seoul. Though dad was a sailor in the navy he was stationed inland because his duties were to the diplomats at the U.S. Embassy located at an army compound near Itaewon. Upon meeting my mom, he married her and adopted me, an infant, as his own, and then bore a son, Solo, before we moved to San Diego. I was a toddler about three years old, and do not remember the move. In fact, I did not know I was adopted until the end of my senior year of high school.

It was a couple weeks before graduation day when Oma told me she wanted to send me to Korea with the youth group from church on a short-term mission trip as a graduation gift. I had to go with my dad to the INS office to get a passport. Off we went, to the lower part of Denver you wouldn't dare go by yourself. It was an area of town where you would find garbage on the sidewalks, the homeless roaming the streets and suspicious characters lurking at every corner. This part of lower downtown was near the train tracks and factories and dilapidated homes sandwiched between Interstate 25 and older skyscrapers creating the northwest skyline of downtown Denver, Colorado! Coincidentally, this is the same part of Denver, which is now called LoDo (an acronym for lower downtown) and a very popular place to live and hang out since the city cleaned it up in the 90's. It is where you will find Coors Field, home of the

Colorado Rockies Baseball Team, The Pepsi Center, Home of the Colorado Avalanche, and the Denver Nuggets. The Union Station, the refurbished home of the light rail, Denver's public transit-rail-system. You will find countless breweries, bars and restaurants fit for the finest foodie! You will still find the homeless, but these days, they are passed by the posh and the popular as some permanent sidewalk fixture, unnoticed. Sad truth.

The year was 1989 and I was about to graduate!

"When 86' is dead and gone,

87' will party on,

88' will think they're cool,

But 89' will always rule!"

This was the mantra I shared with my classmates as we exchanged yearbooks to sign. Reading it today makes me blush. I want to throw it (my yearbook) into a bonfire along with my shameful past. There was one thing about high school I wasn't ashamed of... in fact I was quite proud.

I was on the varsity Lacrosse Team playing Center or Attack positions, and even as the Goalie one season (you should've seen my bruises, lol). I was the player who scored most of the points for my team (when I was not the Goalkeeper). Not trying to brag. It was the truth and my one thing which allowed me bragging rights, because, as you know, my scholastic achievement was a bit wanting. Earlier in the year, I was presented with the chance of a lifetime! A TRY OUT for a full scholarship to Cornell University. My coach and my team elected me to have this opportunity. Cornell was to choose one student from a Colorado high school lacrosse team to join their fellowship. It was my one chance to go to a good college because my GPA alone would not get me there. A week before try outs, during practice after school, I had an accident and twisted my leg. I tore my ACL in my right knee. It was awful, to say the least. Not only was I totally deflated academically by this accident, this was the beginning of the end of my passion to play Lacrosse ever again. My braggadocio days were over.

I don't recall if another teammate tried out for the scholarship in my place, but, my sights were set on Western State College (WSC) after the accident, and I tried to put lacrosse out of my mind as I recovered and watched from the bench as our team lost in the playoffs.

Back then, Denver didn't have as many homeless people as they now do, much like every other major American city, but they were daunting to behold and avoided as the scum of the earth. My dad and I made our way past a brick brown-stone building from a dimly lit parking lot. I remember asking myself why the INS office was open so late at night, and why we had to come to *this office*, located on the opposite side of town from where we lived. Surely there was more than one office!

Through an unseemly single-door entrance from the outside, we walked up a long flight of stairs in a cold and narrow corridor before opening another door to a reception area crowded with foreigners waiting for their name to be called. The room was small, windowless, and not well lit. I felt a wave of claustrophobia rush over me as I looked around for a place to sit. It only stands to reason that we must have had our fair share of waiting in that tiny room, but I have no memory of it. My mind blocked it out, I'm sure, as it was not a very pleasant memory. The next thing I remember is our name being called and we were escorted to a private office.

This office was tiny. Maybe 8x8', no more than 10x10', large enough for one retro-metal desk and chair, and two more chairs opposite the desk against the far wall. As we sat in the two empty chairs I heard a voice from a man I could not see for the files on his desk were piled so high I wouldn't have been able to see him even if I had been standing.

"Do you have her adoption papers?"

I was confused, thinking we entered the wrong office, as I looked over at my dad opening his briefcase and pulling out documents. Before I could ask, my dad stretched out his arm with the papers, just barely lifting himself off the chair, as the INS clerk's hand reached between the piles of files to receive them. The rest of the visit was a blur. I was processing what had just happened.

My dad is not my real dad. *What! Are you kidding me! WTF!* I was in shock as I thought these things on the ride home. The car was silent. Not a peep out of either one of us... the entire way. I don't know what shocked me more, the fact that neither Oma or my dad debriefed me beforehand, or the fact that I was adopted and never had a clue!

Truth be told, I did have a clue. Many clues, but I failed to see them for what they were! The following fall semester as a freshman in the dorms at WSC those clues surfaced as I lied down to sleep night after night, making me sob with sadness. Lucky for me, my assigned roommate came down with MONO and was unable to move in until she recovered. I had the room to myself. High on the hill, on the north end of campus, was Escalante Hall, my new home. I felt free and I was so excited about starting this new chapter of life! Even though the beginning was met with sorrow, I was so elated to be out of the house, and I vowed to never move back home. I was outa there, for real.

The first month I cried myself to sleep as a new *clue* would come to mind. When I was a small child I recalled the joyous occasion of my two brothers going to visit grandma and grandpa, but I couldn't go. I didn't understand why I couldn't go!

Another night as I was winding down to go to bed, another clue would surface. This one, specifically, should have woken me up to the truth because it wasn't a clue at all, it was the plain truth in the wide open.

"Why do you run to him? He's not your daddy!" Oma would say with a scowl when I was still but four years old. You know how toddlers and small children would clutch onto the legs of their parents? I did that a lot, running away from my mom and clutching onto my dad. This must've made Oma furious. Why would she say such a callous thing to an innocent child! I would recall these memories as I lied there alone in the moonlit room, looking out the enormous window adjacent to my bed. As the moon set, so would my mind, lulling myself to sleep on a tear-soaked pillow.

As an adult, more mature in my ways, remembering the remembering now makes me feel petty. Meeting others and reading about others who truly had horrific childhoods, mine was a walk in the park by comparison. At least, I had a family. At least, I had all my needs met. Mostly. Btw, I never did go to Korea after graduation. There were problems issuing me a US passport. It took nearly another 11 ½ years before I finally got one, after having passed the US Citizenship test. I really didn't want to go anyway because I didn't want to miss a special summer lacrosse tournament in Vail. It would be my last.

Amid the chaos in my home growing up, I am so thankful now for the many blessings not withheld from me. I am most grateful to my parents for making me go to church (before I was old enough to resist) where I learned all the Bible stories I now hold near and dear to my heart. I forgave my mother for treating me like a boxing bag, although I must admit I have an ongoing repentant heart toward her. If I do not actively choose to forgive her, bitterness will try to wriggle its way back in! It would not have been possible if it weren't, ironically, for her determination to raise us up with Bible knowledge. It amazes me how God can transcend through the ugliest of situations and reveal HIMSELF even when the circumstances would seemingly point to confusion and chaos. Despite my less-than-ideal home-life the Bible stories instilled in me at a young age anchored me. In all my bad choices and loud surroundings there was a still small voice calling me all the time. This voice was kind and it was patient and loving.

It was an enduring love. It shined more brightly as I ventured down a path of darkness. I've realized it is exceedingly difficult for humans to shine this kind of love. It takes concentration. Everyone is so distracted. God's love is getting harder to find... in people. But His love was ever-present in my life, in spite of it. Impressionable at a very young age, I encountered Jesus and He never left my side, even though I ran from him all the time. His love would find me regardless how far and fast I would run away. He would come as a gentle breeze when I needed the cool of His love as I became overheated by the trials of life. Other times He would come as a cosmic hurricane or an erupting volcano to melt my heart growing cold from rejection or unforgiveness. I have so many regrets from bad choices, but it did not stop His voice from calling to me. In fact, many times it saved me from a horrible fate.

During my 2nd year in college, a handful of my girlfriends and I were to stay in a cabin near Irwin Lake off Kebler Pass for a weekend of snowmobiling. This small cabin was off the grid, and Kebler Pass, itself, was closed to automobiles in the winter, which helped preserve the unspoiled wilderness from commercial access. We had to bring water. And use an outhouse! It was in the heart of winter and during a snowstorm when all of us geared up and packed our snowmobiles for the journey to the cabin from Crested Butte. It was supposed to be a nice respite from our usual skiing/snowboarding-every chance-we-got between classes and every weekend routine, but all the fresh-falling snow made me long to ski instead. We stopped at the Bakery Café before we headed to the snowmobile outfitters. It was already sundown when we huddled together as a group to hear Kris give us the rules before we made our way to the cabin.

"Stay close but not too close. Before we get to the path to Irwin, we will veer right to the cabin. We must stick together so no one gets lost," Kris would say in a commanding tone. She was a friend from high school, but here and now she was our guide. She, in fact, was a professional snowmobile guide in the winter and river-raft guide in the summer. She was in her wheelhouse and on her soapbox. She was qualified. The snowmobiles we were using belonged to her boyfriend who owned a snowmobile excursion company in Crested Butte, and the cabin we were going to, he built with his hands, I think.

I remember *my hands* were freezing and cramping up from constant applied pressure on the throttle. Many times, I wanted to stop, but couldn't. There was no stopping. To the cabin or bust! A couple times I had to remove my goggles to defog with one hand, hoping I would not accidentally drop them. The snow was coming down hard. I was glad to be with a group led by Kris because it was a little unnerving being this far from town in the cold dead of night. In a blizzard, no less! I kept thinking to myself, tomorrow would be an epic powder day on the mountain. The *ski* Mountain. Crested Butte Mountain. Behind us. But we're stuck on this one. We are going in the wrong direction! I was a skier, not a snowmobiler. I had to resist the temptation of bumming out because I was going to miss what would probably have been considered the best powder day of the year!

Dang it! We should've brought our skis and boards! We're in the backcountry, for God sakes! We could take turns towing each other up with the snowmobiles and ride down. Argh! Too late. I'd shrug off those thoughts which kept returning. No sense in dwelling on that I kept telling myself. We had this planned out for weeks. We will have an awesome weekend snowmobiling!

The cabin, a simple A-frame, was covered in snow and nestled in a small clearing in a small valley surrounded by snow-covered trees. It was a truly magical sight, yet familiar; one I had seen before in a Thomas Kinkade painting. The falling snow, as thick as it was, made it hard to see when we were traveling on the pine-tree-lined trail, but you could still see the warm incandescent lights shining through the windows as we approached it and parked our snowmobiles in single file near the front porch. It was quaint inside with a living/family room and a roaring fire already in the stove, and the aroma of Patchouli oil in the air. It was cozy, and definitely lived-in with all the personal touches decorating the small place. I think Kris had moved in and she and her boyfriend made it their primary residence. The modest kitchen (with no running water) was opposite the ladder and wood stove in the center of the room. The ladder led to our sleeping quarters. A loft large enough for all of us to snuggle together within our own mummy-sleeping bags. *I can't remember where the bathroom was… probably a freezing stroll away from the cabin, and my brain blocked out that unpleasant memory. Lol!

They had a pet chinchilla free to roam the house and I recall trying to find it the whole time I was there. I think we all were searching for this elusive animal. We just wanted to pet it! It was like an all-night game of hide and seek. We never found it.

The next day, after a pancake breakfast, we made our way to Irwin Lake where Irwin Lodge used to be… a go-to retreat for highfalutin-city-folk longing for an authentic Colorado Mountain Getaway. I don't believe it is still operational today. *I'd love to go see if it's still there and what they made of it!

Just before leaving for our day of snowmobiling, Kris stepped up on her soapbox and gave us the rules.

"Stay in file until we get to the lake. The lake is frozen, and we will spend the day playing on the lake. DO NOT venture off on your own. There are many paths on the outer banks but be sure to stay on the lake at all times." And off we went.

It had stopped snowing, but the day remained gray and looming over us were thick dark clouds which seemed to say more snow was on the way! When we arrived at the lake everyone seemed to twist their throttle, gunning it, and scattering into all directions of the wide-open space. You could see Irwin Lodge at the far side of the lake aloft an escarpment. The lake was clear of tourists and locals. We had it all to ourselves.

At about noon we all came together to eat our sack lunches. Afterward, Kris reminded us of the rules and gave us the time we were to all meet up again before heading back to the cabin at the end of the day. That afternoon as I was exploring the outer banks of the lake, I saw a path which was so inviting I could not resist to take it. The lake, itself, was getting boring and, although I knew the rules, I convinced myself there would be no harm in me taking this trail. The path was wide and smooth, so I gunned the throttle! I was moving as fast as the snowmobile would go to compensate for the elevation I was climbing and the foot (or more) of new snow on the ground. Up and up; this trail led beyond the lodge.

The path began to wind slightly so I let off the throttle, but it was too late! I nosedived into a creek crossing the trail. I came to a slamming stop, the snowmobile perfectly vertical, its front skis wedged between two boulders from what I could tell as the snow exploded around the impact zone. The creek was frozen and a good four or five feet drop below the path. I was in a trench! Upon crashing, my body hurled into the front of the beast of the machine I was driving. My right cheek was smashed against the windshield, my arms twisted behind my torso, and after God-knows-how-long, as I shook off the paralysis of the shock, I tried to move.

"Okay, my arms seem to be okay," I said to myself, as I lifted the weight of my body off the handlebar. "Aaaaah ooouuuuuuccchhh!" I jolted as I screamed from the pain now radiating from my left knee. My GOOD knee!

"Oh no!" I cried, "this can't be happening!" I had to take special care moving my left leg, which wasn't easy considering the position of my body and the gravity of this vertical situation I was in (pun, hehe). Apparently, my left knee took the brunt of the impact. It felt crushed! I slowly and meticulously lifted my left leg, picking it up by grabbing my thigh with my left hand. I had to keep my right hand on the machine to prop me up because I was in a near-somersault position. My right arm was starting to burn from the payload... of me!

Wait a minute, what of my other leg, I thought to myself. I moved it and it was fine.

"Thank God," a sigh of relief followed. I swung my good knee (actually, my bad knee, lol), which in this situation was my good knee... I swung it around and found footing above the side rail. As I shifted the weight of my body I freed up my right hand to assist my left hand so it could painstakingly lift my entire leg over the saddle of the machine to the right side, so my good "bad" leg could bear the weight of my next move.

Come on Gwyn, you got this, I would encourage myself silently, as I gently rested my left leg over the opposite side of the seat parallel to my other leg. Now that I was on the right side, I arched my back over the saddle as I deliberately held onto my left leg to keep it from moving, moved my right leg, and allowed myself to slide off the seat and onto the boulder beneath it.

"Aaaahhh oooouuuuuccccchhh," I shrieked as I landed on the rock with a thump! I managed to crawl out of the trench and find a sturdy broken tree branch to use as a crutch. As I regained some composure, I assessed my situation and realized the path I was on took a sharp turn to the left just before the steep embankment of the creek. They did not cross paths, after all, but who could tell with all the snow and the flat light conditions! The velocity at the time I met this curve didn't help either! I was so glad I did let off the throttle beforehand because I had the pedal to the metal, as they say. It could have been worse.

I looked around; back down the path I had just climbed, but I could not see the lake for the trees. I began moving down the trail but quickly realized I had to make my way to higher ground. I followed the path I was on, limping, as I favored my good "bad" knee. I was moving slowly, but rhythmically. There was a clearing of trees around the bend, so I headed there. I hiked up the trail, made more difficult by the deep snow. At least below, I could stay on the track the snowmobile made as it compacted the snow and made the path easier to tread. Here, moving up was especially exhausting, but I was making headway. My cries followed the beat of the drum; I'd howl as the sharp pain in my knee would spring anew with every grueling step. It was as if I was beat boxing! To a slow song.

I found a stump to rest on when I reached the clearing, and from there I could see the lake. I shrugged, took off a glove and wiped my eyes. Holy cow it was far away. I had wished I had binoculars. More so I had wished I had a two-way radio.

I lifted my arm and put my hand in front of my face, squeezed my forefinger and thumb together just enough to use it as a window through which to peep. Between my finger and thumb was no more than a quarter inch of space; just enough to fit the snowmobiles moving below on the frozen lake.

"They look like ants!" I exclaimed, "Damn it, damn it, that's far!" I screamed for help as loud as I could, knowing there was no way they could hear me.

"How did I get this far off the lake? Why didn't I follow the rules!" I snarled, as I complained to myself beating the cold air with my fists causing my whole body to shake. I shook myself as sheer panic began to surface, and I started to hyperventilate.

"WTF!, f**k, f**k, f**k!!!" I would shout as loud as I could. Then I would rupture into tears and growls and screams. Then I would start to hyperventilate again and punch the air now growing colder.

Over and over again, this was all I could muster!

"Oh my God, what am I going to do!" I squalled. *Will I survive the night if they don't find me?*

*The road to my soul may have on-ramps everywhere, but I can say with confidence today that many of them (and numbers still growing) are met with roadblocks! These roadblocks are not your typical concrete barricades. Oh no! The blockades I am referring to are guards. Guardian angels! They stand guard at the gates of my heart! Oh, and by the way, sunglasses are no longer necessary, for those in Christ have new eyes to behold His Glory!

Chapter F1VΣ

Challenged

Gunnison, CO sits at an elevation about a mile and a half above sea level and is in the middle of the mountains, but the town itself sits on a plateau, "the "valley," which is fairly flat considering it is in the heart of the Colorado Rockies! The campus of Western State College (WSC) takes up the east side of town and is situated on a flat area at the front of the campus but climbs hills as you move to the back of the campus. Escalante Hall was at the far back end. I got a lot of exercise having to go uphill each time I headed to my dorm, but it was worth it. I had a nice view of town and the Palisade Cliffs westside of the plateau. The Common Area at the center of Escalante Hall was poised for sunset viewing and had the best view on campus and *of* campus... a view of the entire campus, itself, and the whole Gunnison Valley, as well as the huge W monogram on the hill (or mini-mountain) opposite the county road entering Gunnison from the east. The W was there to remind us all of school pride. *Yeah, Wasted State... so proud.* You could see the front side of Hartman's Rocks and you could even see the San Juan Mountain Range far to the south. Unlike the south where the space is wide open, Gunnison to the north narrows as it follows the Gunnison River upstream on County Road 135, just a few miles, going past a small canyon where the river begins at Three Forks, from the Taylor and East Rivers.

If you followed the Taylor River it would take you through Taylor Canyon heading northeast and up to Taylor Reservoir, a stunning alpine lake, west of Cottonwood Pass from Buena Vista. If you followed the East River it would take you north northwest to the ski town of Crested Butte, known back then for its fresh powder and still today for its wildflowers... in summertime, of course. This entire area, Gunnison, Three Forks and Taylor, as well as Crested Butte, is one of the most beautiful places on God's Green Earth. Oh! Blue Mesa and the Black Canyon (to the west of Gunnison) too! Gorgeous!

Most students return home during summer break and I had too, until this summer. Gunnison had so much to offer and I had only previously experienced the wintertime offerings. The skiing, the ice skating, the snowmobiling! I had really begun to take mountain biking more seriously and wanted to spend the summer exploring all the trails in Gunnison and Crested Butte. My roommate, Heather, decided to stay for the summer and so did I. She and I were roommates our sophomore year while living on campus in campus apartments for upperclassman, but now we were living off campus in a single-family home made into a duplex on North Pine Street like "real" upperclassman. Heather and I lived in the back of the house which was converted to a two-story unit from what was once the garage. It was luxurious compared to our campus apartment. We each had our own bedroom, and we had 1 and ½ baths! We had a full kitchen, and a shared laundry room with the housemates in the front of the house. We had it all! We even had our own cars! We really were upperclassman now!

This day, instead of hitting the dirt trails, we decided to do something different. We bought a two-man raft, The Challenger 2000, from Walmart.

Walmart had just opened its doors in Gunnison, after many locals protested to no avail. Students, like us, loved it for its low prices and numerous options to buy stuff, but the owners of local ma and pop shops were not too happy to see this monstrosity of a building be erected so close to town. In town. On the north end.

Together, Heather and I managed to haul the box home, unpack it, fill it with air and load it up on the ski rack atop her car. We headed to Taylor Canyon. That is, after stopping for gas at the station which used to be the last commercial building on the north side of town, except for the dated movie theater farther up on the west side of the road. Now, the last building on this side of the road and on this side of town was Walmart.

While pumping gas I was thankful for having a car as I reminisced the two years before having been at this same gas station, in line on the side of the road, hitching for a ride to Mount Crested Butte. Or "the Butte" as we called it.

This was the official hitchhiking "depot" for students who needed a lift to the ski resort. It was the norm and the thought of danger never entered our minds. Until one day. A friend, Angie, and I were in line waiting for our turn to stick out our thumbs. When our turn arrived, we didn't have to pose our thumbs because a VW Rabbit had already pulled over. This was a known spot and upperclassman, who once stood here themselves, like me, were graciously able and willing to give lower classman a lift to the mountain. Professors, too, were known to give a lift on occasion, and some on a regular basis. I was entranced in the memory.

"I only have room for one because I have to stop and pick up a couple more friends," the driver said after she stretched over rolling down the passenger window manually, from her super-cute baby blue VW Rabbit.

We let a person behind us take this ride as we did not want to go separately. We waited for the next opportunity and it came immediately.

A beater of a car pulled over and a man weathered with decades of winter storms rolled down his window and smiled at us and he looked us up and down with no shame. The deep-set wrinkles on his face were more deepened and darkened as he smirked. His smile widening; he could barely keep in his tongue.

"Got plenty of room for the both of ya!" He exclaimed. We felt like mannequins in a display window at Macy's.

"No thanks, we're waiting for a friend," we said speaking over each other, trying to find the right words. We were the only ones left in line, so this creepy guy drove off alone. Fortunately. We could see his shotgun resting behind the backseat from the back window, as he switched gears and merged onto the road and left us in his dust. Angie and I looked at each other with wide eyes and a look telling we had the same thought. Had we gotten into that car it may have been the last time anyone ever saw or heard from us again. Shaking off that memory, I was now so thankful to be living here, in Gunnison. Attending *Wasted State*. I was happy with my life. *I actually might be somebody,* I thought as I softened a smile.

I finished up pumping gas and went inside to pay. *Yeah, back then we didn't have the option to pay with a credit card at the pump." At least not at this station. Heather was driving ahead of me with the raft vibrating from the wind. I hope we tied it down well enough, I thought to myself as I watched it rattle. Making our way past the Lost Canyon and the curves headed north on the 135 just beyond Almont Campground, we had to slow down to turn right onto 742 toward Taylor Reservoir. This is the fork where the three rivers meet, The East and Taylor Rivers join the Gunnison River. Creating it. As we slowed through the intersection, I looked at the Three Rivers Resort, where Kris (remember, my friend and snowmobile guide) worked as a white-water raft guide in the summer. She was probably there, on her soap box, giving the rules to the bubbling group of tourists, itching to ride the rapids of the Taylor.

From Three Forks we went upriver about three or four miles to drop off my car at the TAKEOUT site. I jumped in Heather's car and we headed up another four or five miles, near Harmel's Retreat, to the PUT-IN site where we would begin our adventure, on the *Lower Taylor*.

We geared up and hopped in the raft, I took the front, she took the aft! *Ooh, I should turn this into a poem!

The Taylor River begins at the reservoir 26 miles up the canyon from Three Forks. As opposed to the *Upper Taylor,* the Lower Taylor was more moderate when it came to waves. We felt confident The Challenger 2000 would be up for the challenge. From the get-go we were paddling hard to avoid rocks and boulders, and to steer the raft. We realized pretty quickly that steering this thing was not so easy.

"There are a lot of rocks!" Heather yelled, making sure I could hear her over the white noise of the rushing river. The water was moving fast... too fast for this raft. The water was cold. We were in over our heads! The waves were huge, at least they looked that way from this tiny vessel we were riding. Less than halfway down our planned route there was a POP! The Challenger turned into a deflated weasel within seconds. We were sunk and we were swimming. Fighting the rapids, it's a wonder we were able to hang onto the cumbersome and now made heavy vinyl raft. We dragged our sorry asses up a rocky bank along with the sorry ass raft. Totally soaked and deflated ourselves, we looked at each other and started laughing uncontrollably. Belly laughs.

Sitting on the bank, catching our breaths, we exchanged reasons why our raft popped.

"We shoulda steered left of the rock,"

"No, we shoulda gone right of the fall!"

In the end we both agreed we got what we paid for!

There was a small embankment we had to get over in order to get to the road. We would hitchhike from there. Hopefully, it wouldn't be a problem due to the fact we were soaking wet! As we made our way over the mound all we could see was another embankment. As we cleared that mound we saw yet another embankment farther up, but this time we could see that we were nowhere near the road. For the most part, the Taylor River hugs the road for the entire 26 miles up the canyon, except for one part on the Lower Taylor. There's a small stretch where the river and road move away from each other in an open valley separated by farmland. That's where we found ourselves!

Out of all places to be stranded this is where we ended up; where the river is nowhere near the road and along a pasture so big you couldn't even see the road! Jeez-o-pete! It wouldn't have been so bad had we not had a raft to drag. We tried to fold it and make it more manageable to carry. One of us had to hold the valve open as the other rolled the now waterlogged raft. It was more work than it was worth. Holding the valve open was painful to stiff cold hands. We couldn't get the raft to fold, partly because we couldn't expel the excess water and air and because the raft was so stiff from the freezing alpine river and cold mountain air. Even though it was summer, at this elevation the air stays chilly, especially in the shade. The river, on both sides was nestled by evergreen trees. As short-lived as our adventure was, it was fun. But it was cold!

Shivering, we moved closer to the third embankment and we cleared the forest of trees. We could hear what sounded like thunder. A low rumble. As we cleared the third embankment we had to jump a split rail fence wrapped in barbed wire and toss the raft over, not caring if it snagged. We weren't saving the Challenger in order to repair it and reuse it. We wanted our money back! Although it was a cheap toy, every penny was worth a hundred bucks to us college students!

There was yet another embankment to cross, well, it was more of a slight incline. We continued slowly; dragging the Challenger caused a lot of drag. And was a drag! We realized the thunder was getting louder and the ground was shaking. Once we could see over the top of the incline both Heather and I, in unison, dropped the raft like a hot potato and screamed, "SHHHHIIIIIIT!" as we turned around and ran back toward the river, hopping the fence like trained gymnasts and tumbling to a stop on the other side just as a stampede of horses brushed and rushed by the fence following it to the south down river to a watering hole!

Perched on the ground, at the ready to make a run for it as if our lives depended on it we were frozen. Staring at the beautiful creatures galloping by, we were mesmerized by their impressive speed and stamina. They were so close. It was like having front row seats at the Kentucky Derby. Better yet, like bench seats on the track itself! Their manes and tails were gorgeous, colorful, and long and glistening and as if flowing in slow motion. I thought I was dreaming. They were moving as fast as the river behind us! One second they were there, the next they were gone. Like the wind. It was a close call. The ground where they passed was now covered in hundreds of pock marks stamped by their hooves. You could barely see the raft, now partially buried in the earth, and pasted to the upturned mud. It was now flat as a pancake. And probably foldable! Holy cow, that could've been us! After our hearts settled back in our chests, Heather and I turned to each other in total shock, and burst out with a second round of belly laughter.

$$G + H \times C / T R^2 = TB$$

Gwyn plus Heather times The Challenger divided by the Taylor River's Rapids squared equals a TOTAL BUST

*Okay, as I skimmed through my file of poetry, look what I found! Written in 1993. What a truly delightful surprise. I belly laughed. I had written a poem, after all...

The Challenger Challenge

Struggling in shallow waters
Because the Challenger slipped
Flipped over a rock!

Shocked by the freezing rapid
Trying to clutch the current
Only kissing the swell!

The water was too fast
To sustain oneself
And to shallow to swim!

Supple to waves
Bruised by rocks
Scraped by the wet floor!

A dreary moment
Trying to clutch the current
Only kissing the swell!

Screaming with water-filled lungs
The Challenger failed
Gyrating to the jive of waves!

1993

*So, after reading this, Heather and I must've flipped over a rock! And that's what caused the raft to pop. Wow! I didn't recall how bad it really was... In fact, I don't remember anything after the horses. It's all a blur from there. I don't remember if we crossed the field or walked around it. I don't remember hitchhiking back to the car, and I don't recall if we ever returned the raft to Walmart. For all I know it's still there smashed to the ground, or buried, near the fence along the river. But I will never forget the horses, the look on their long faces and the flow of their manes. The horses won the day, and, so it seems, that part of my long-term memory! Not long after the Taylor River incident we decided to boat again along the Gunnison River on a very mild stretch from town center to the west. For me, that was my last routine on the river because I had another terrible experience. Not with the rapids but with the tiny leeches. Ick! We still ventured on annual multi-day trips to Westwater Canyon, Utah (big rapids on the Colorado River). Oh, and I did the Goosenecks (class one float trip) on the San Juan River, but no more regular local river rides for me. For Heather, she bought a squirt boat (kayak) and hit all the rapids at all nearby rivers all the time. She was fearless, until one day, years later, on the Arkansas River, she got stuck in a "pourover" and was pulled beneath the wave. She gave up kayaking after that harrowing brush with death. From what I understand, she has now begun again, decades later.

God *challenged* me my whole life. I failed most of the time. I have a long history of failures. The ones I regret the most are the ones I lost to God. I wish I could go back in time to fix them. How many chances do we get? How many times do we long for a do-over? If only I could go back in time and undo all my wrongs. If only I could forget. Sometimes, a rare opportunity will present itself and a chance at redemption is at hand. They are hard to recognize when our eyes are downcast. Looking upward we see beauty and hope for a bright future. Even when we're stuck in the mud we can, at least, turn our attention to the gorgeous horses galloping by. We can catch a glimpse of a movement unadulterated by our circumstances.

There is something out there. Something bigger than myself. Something glorious and powerful. Something that will move along and take its course in spite of me. Something that will move past me whether or not I choose to look at it.

Surely, we've all experienced this. Many of us will ask: Will it stop for me? Will it call my name? Does it know me?

Bowel Movement

Sticks and bones trapped under stones
Years and years and pressure more
Chaffed by rocky teeth and swallowed into the stomach of Earth
Digesting matter of fat, feces, flowers once fresh, dead flesh, all things dead
trees, teeth, seeds, and weeds
A diet contributing to her nourishment indeed
Root rot too serves nutrients to her malnourished ground

Fertilize the dermal layer for fuel for heat respectfully
Metabolic magma, pumped by her heart,
raging through arteries internally
Time to time the colon of Earth collides with a vein,
but normally maintains
a safe passage to the bowels
The foul remnants are swallowed for nutritious topsoil
And some are excreted as nothing more than oil
Everything digests by the bile of the body of Earth

Then therein defecates
soil and oil in pools of stool
As we recycle our excrements to her soiled skin
She generously excretes her recycled share back to us
out and in the crust within

Us to Earth and Earth to us

1996

Chapter S1X

Absolute Zero

absolute

[a b-suh-loot, ab-suh-**loot**]*adjective* free from imperfection; complete; perfect: *absolute liberty.*

not mixed or adulterated; pure: ... complete; outright:...

free from restriction or limitation; not limited in any way: *absolute command; absolute freedom.* see Dictionary.com

It's January 2020 (as I write this), and I've heard people say as they greeted each other and wished well of each other in the new year, that we are in a new decade. Words of wisdom given from sermons and speeches... blessings for resolutions I've heard in prayers, and annual predictions on the news; from various sources they all agreed we are beginning again.

Happy New Year! Can you believe it! We made it! Everyone seems to think we have entered into a new decade.

But have we? Wouldn't 2020 be the last year of the current decade, and the new decade begin with 2021? After all when we count we do not start with zero. We start with one and end with 10, in as far as our fingers demonstrate. Once we reach 10, we start over. We, thus, begin anew with 11, or 21, or 31 or 301, or 2001. I can see why people would think we've entered a new decade; because the numeral representing the next decade precedes the zero, giving the illusion that we have entered a new ten-year cycle. We are out of the one's and entering the two's, from all appearances.

It all starts with one. Or does it? A new day and the first hour begin at zero seconds and counting past midnight and the 2nd hour begins at 1:00 am. 2:00 am starts the 3rd hour. Can you imagine how confusing it was in ancient times using sundials and other clock relics? At what hour do you begin?

When do you turn over the hourglass for the sands of time to start again? There is a real paradox here, and it all starts with the tracking of time.

The Julian Calendar, was established by and named after Julius Caesar when he reformed the Roman lunar calendar to join the Egyptian solar calendar in 45 or 46 B.C. after meeting Cleopatra and restoring her throne to Egypt as co-ruler with her brother Ptolemy XIII (not to be confused with the Greek mathematician and astronomer Claudius Ptolemy).

In order for the Romans to reconcile their calendar to the solar calendar they had to add two extra months. History tells us 46 B.C. was 445 days long. To account for the imperfection of their timekeeping they added those extra days to stay in harmony with the seasons as ruled by the sun. The Egyptian solar calendar was not perfect even though they were the first to have 12 months and 365 days in a year. More accurately, they had three 120-day seasons and added the extra five days at the end of the last season. They celebrated during these extra (five) days at the end of their year as festivals. Then they would start the count over again with their new year. The Romans decided, instead, to add leap years to account for the extra days, adding a day to Februarius every four years.

A true solar year, we know now, is 365.242 days long. That is a very complicated number, and even with leap years or five extra days, their calculations would reveal errors and the .242 margin would catch up to them and not line up with the sun. The farmers of the day were getting confused as when to plant their grain. Lol!

The Egyptians in their effort to account for the extra days still would come up a quarter day short every year and would have a whole extra day after four years, which came to be known as their Wandering Year. The Julian Calendar with its leap years came up short too, whereby in 1582 the Gregorian calendar, named after Pope Gregory XIII, took its place and is the calendar we use globally today. It is essentially the same calendar with minor changes. The only difference is the Gregorian calendar adopted a finer formula... allow Wikipedia to explain:

> The calendar spaces leap years to make the average year 365.2425 days long, approximating the 365.2422-day tropical year that is determined by the Earth's revolution around the Sun. The rule for leap years is: Every year that is exactly divisible by four is a leap year, except for years that are exactly divisible by 100, but these centurial years are leap years if they are exactly divisible by 400. For example, the years 1700, 1800, and 1900 are not leap years, but the years 1600 and 2000 are. The calendar was developed as a correction to the Julian calendar, shortening the average year by 0.0075 days to stop the drift of the calendar with respect to the equinoxes. To deal with the 10 days'

difference (between calendar and reality) that this drift had already reached, the date was advanced so that 4 October 1582 was followed by 15 October 1582. There was no discontinuity in the cycle of weekdays or of the *Anno Domini* calendar era. The reform also altered the lunar cycle used by the Church to calculate the date for Easter (computus), restoring it to the time of the year as originally celebrated by the early Church.
https://en.wikipedia.org/wiki/Gregorian_calendar

Furthermore, Wikipedia does a great job explaining how the Gregorian Calendar became established worldwide as our modern-day choice to track time, fyi:

The reform was adopted initially by the Catholic countries of Europe and their overseas possessions. Over the next three centuries, the Protestant and Eastern Orthodox countries also moved to what they called the *Improved calendar*, with Greece being the last European country to adopt the calendar in 1923. To unambiguously specify a date during the transition period, (or in history texts), dual dating is sometimes used to specify both Old Style and New Style dates (abbreviated as O.S and N.S. respectively). Due to globalization in the 20th century, the calendar has also been adopted by most non-Western countries for civil purposes. The calendar era carries the alternative secular name of "Common Era"
https://en.wikipedia.org/wiki/Gregorian_calendar

In the book, Zero, by Charles Seife, he writes that the Julian Date was an integral factor in the new Gregorian calendar established by Pope Gregory XIII in 1582 (the Julian Date should not be confused with the Julian calendar... the Julian Date was actually named after Julius Caesar

Scaliger, an Italian physician in 1583 (pp 58)). His son, Joseph Scaliger, a scholar and known for being the father of Chronology (see https://www.lindahall-.org/joseph-justus-scaliger/) integrated the tracking of time based on astronomy in helping to establish the current Gregorian Calendar. The Julian Date follows astronomical events starting at January 1, 4713 BC when three multi-year cycles began. *I'm still trying to figure that one out… it's rather confusing why they chose that date as their start point. Regardless, it is still used today by astronomers worldwide. Although it seems only minor changes were made in the transition of Julian to Gregorian, the Gregorian took into consideration calendars from all nations, but more importantly, astronomical events. But even the Julian date was modified since then.

I wouldn't be surprised if modifications would have to be made again in the future in order to synchronize our clock to the grandfather of all clocks. The Universe. If the universe is indeed expanding, we would have to account for the time differential as it would, no doubt, have an effect on our seasons. The solstices and equinoxes determine our seasons, but they are determined by the seasons of our solar system, and our solar system is determined by the seasons of our Milky Way Galaxy! And so on.

Back to the calendar, as we know it. Just like our 24-hour day begins at zero seconds and counting after midnight, our years do the same. Let's start at the beginning of A.D. On 1 A.D., the very first day of the first millennium, the calendar clock began, or did it?

The Roman Catholic monk, Dionysius Exiguus (not to be confused with the Greek god Dionysus), on order of the pope, Pope John I, created an extension of the Easter Table. He had to recalculate the days going back to the birth of Christ, and in so doing he established Anno Domini, Latin for Year of Our Lord. This is where we get the term A.D. in reference to the calendar. Most people, including myself, believed that A.D. stood for After Death, which is logical considering B.C. stands for Before Christ. Anno

Domini was, thence, used ever since until recently. The count began with the day of our Lord, in which Dionysius calculated using what was available to him. Dionysius hence made an adjustment to the already established Julian calendar in order to accommodate for keeping with Easter. This is one example of the Roman Catholic Church adjusting their Julian calendar every couple hundred years (estimating here) or so when they realized that the table they used to count for Easter was off. They measured the discrepancy from the first full moon after the spring equinox. In order to stay true to Easter they would extend the table making adjustments accordingly.

Dionysius made an error for the first day, 1 anno Domini (1 AD), because it was later found and universally agreed upon that Jesus was born four years earlier, based on the death of Herod The Great which gave a reference point for the timeline of Jesus' birth. It was later determined and confirmed that he indeed was off about four years, using astronomical events, via the Julian Date I imagine. The Christmas Star was a triple conjunction of planets, and then there was a lunar eclipse recorded at the time of Herod's death. These findings led some people to believe that in 1996 Jesus was 2000 years old. What! We are four years behind? This error was not adjusted with the adoption of the new Gregorian calendar. As we read above from Wikipedia, the only adjustment made was for nine days; to line up with the equinox. Apparently, there was never a correction made to Dionysius' error. To make matters worse, there is another variable unaccounted for... moving from BC into AD... counting backward... then forward...

Author, Charles Seife, illustrates in his book, Zero, why we are in the 21st century and not the 20th. Why the 1700's is considered the 18th century, and the 1900's is the 20th, etc.:

Imagine a child born on January 1 in the year 4 BC. In 3 BC he turns one year old, and in 2 BC he turns two years old.

In 1 BC he turns three years old. In 1 AD he turns four years old, and in 2 AD he turns five years old. On January 1 in 2 AD, how many years has it been since he was born? Five years, obviously. But this isn't what you get if you subtract the years: 2- (-4) = 6 years old. You get the wrong answer because there is no year zero.

By rights, the child should have turned four years old on January 1 in the year 0 AD, five in 1 AD, and six in 2 AD. Then all the numbers would come out right and figuring out the child's age would be a simple matter of subtracting -4 from 2. But it isn't so. You've got to subtract an additional year from the total to get the right answer. Hence, Jesus was not 2000 years old in 1996; he was only 1,999. It's very confusing and it gets worse.

Imagine a child born in the first second of the first day of the first year: January 1 in one AD. In the year 2 he would be one year old, in the year 3 he would be two, and so forth; in the year 99 he'd be 98 years old, and in the year 100 he'd be 99 years old. Now imagine that this child is named Century. The century {Century} is only 99 years old in the year 100, and only celebrates its hundredth birthday on January 1 in the year 101. Thus, the second century begins in the year 101. Likewise, the third century begins in the year 201, and the 20th century begins in the year 1901. This means that the twenty-first century - and the third millennium - begins in the year 2001... 2

He explains that until the number zero was accepted, the ordinality verses the cardinality of numbers were interchangeable (pp 59). The value of the number being cardinal and the order of the number being ordinal.

Counting beginning with 0, 1, 2, 3... is unnatural, whereas, counting backwards from ten to zero is not. *Interesting, huh? So, when did zero enter the scene?

In ancient times numbers were only necessary when people had to count things used in everyday life, and zero had no place in the scheme of things. Tally marks were the most common, much like the Roman numerals I, II, and III. These glyphs were found worldwide and makes sense considering each tally mark represented 1, 2, 3, etc. But there's only so many tallies you can make before it gets ridiculous. Romans used letters/symbols beyond three tally marks. Other civilizations used other symbols. They would begin to group numbers with a single symbol when the counting reached large numbers. It makes sense to group numbers in fives or tens because of the number of fingers on our hands made it easy to understand. Many counting systems, therefore, used a quinary method. However, ancient nations used many different systems; binary (counting in groups of two's (which we still do today in computer computations for digital devices, and for more compact notations computer programmers use base eight and 16)), even 12, 20 and 60 base systems, but, the most common was a base 10 number system. Assigning higher numbers in groups of 10 makes sense to me! For fun, I have included numeral systems so we can see the ancient variations from different cultures and nations.

Figure 12. Roman numerals

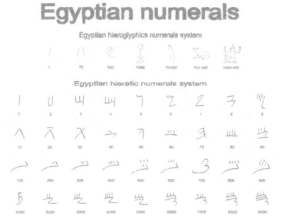

Figure 13. Ancient Egyptian numerals

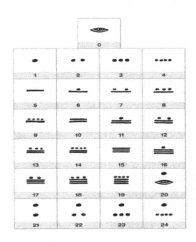

Figure 14. Mayan numerals. Notice they bring their zero as a place holder in their vigesimal (base-20) system

Figure 15. The Mayans pictorial numeral system, Maya Head Numerals. Can you imagine solving math problems with these!

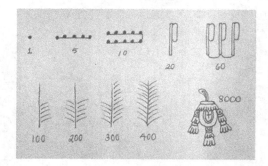

Figure 16. Here you can clearly see Aztec sets of 20. Mayans and Aztecs both used a vigesimal system (base 20)

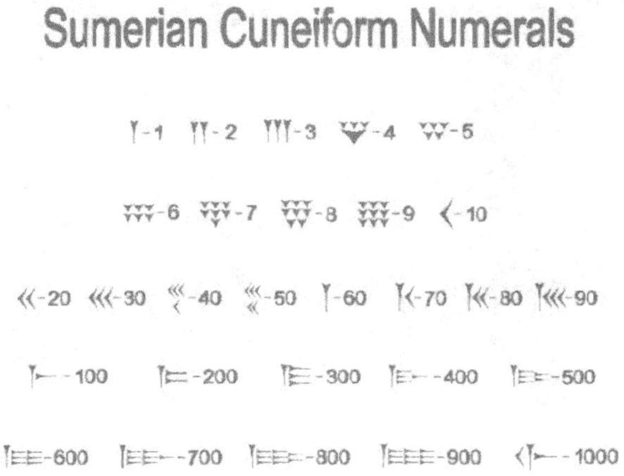

Sumerian Cuneiform Numerals

𒁹-1 𒈫-2 𒐈-3 𒐉-4 𒐊-5

𒐋-6 𒐌-7 𒐍-8 𒐎-9 𒌋-10

𒌍-20 𒌍-30 𒃻-40 𒃻-50 𒐏-60 𒑂-70 𒑃-80 𒑄-90

𒐘-100 𒐙-200 𒐚-300 𒐛-400 𒐜-500

𒐝-600 𒐞-700 𒐟-800 𒐠-900 𒐡-1000

Figure 17. Sumerian/Babylonian Numerals

Figure 18. Sumerian cuneiform stone tablet

Figure 19. Nippur, Old Babylon. The excavation site of Sumerian cuneiform tablets

Babylonians and Sumerians used a sexagesimal system. Notice the typical chevron shapes found in their writing. These wedges were commonplace in tablet writing which granted the name cuneiform (which means wedge in Latin) found in all Mesopotamian tablets including Sumerian, Babylonian, Akkadian, Assyrian, Hittite, Persian, etc. In 2010, 13 cuneiform tablets were exhibited at New York University's Institute for the Study of the Ancient World. If I understand correctly (from the following NY Times publication) some of the tablets featured were done by students copying the ancient Mesopotamian technique of stamping reed into [malleable] clay. Reportedly, the students learned to speak a Semitic language (Akkadian) and were learning to master Sumerian mathematics using their sexagesimal system. Many original Sumerian tablets were excavated from Old Babylon. Nippur is located in Iraq, between the Tigris and Euphrates Rivers, where the Biblical navel of mankind can be found. Nippur is considered the scribal training epicenter of ancient Mesopotamia. The following excerpt is taken from the article, I mentioned above, which announced the tablets on display at NYU on November 22, 2010 by Nicholas Wade, a British author and scientific journalist for the New York Times's Science Times. It's entitled An Exhibition That Gets to the (Square) Root of Sumerian Math:

> ...Sumerian math was a sexagesimal system, meaning it was based on the number 60. The system "is striking for its originality and simplicity," the mathematician Duncan J. Melville of St. Lawrence University, in Canton, N.Y., said at a symposium observing the opening of the exhibition.
>
> A 59 x 59 multiplication table might not seem simple, and indeed is far too large to memorize, so tablets were needed to provide essential look-up tables. But cuneiform numbers are simple to write because each is a combination of only two symbols, those for 1 and 10.
>
> Why the Sumerians picked 60 as the base of their numbering system is not known for sure. The idea seems to have developed from an earlier, more complex system known from 3200 B.C. in which the positions in a number alternated between 6 and 10 as bases. For a system that might seem even more deranged, if it weren't so familiar, consider this way of measuring length with four entirely different bases: 12 little units, called inches, make a foot, 3 feet make a yard, and 1,760 yards make a mile.
>
> Over a thousand years, the Sumerian alternating-base method was simplified into the sexagesimal system, with the same symbol standing for 1 or 60 or 3,600, depending on its place in the number, Dr. Melville said, just as 1 in the decimal system denotes 1, 10 or 100, depending on its place.

The system was later adopted by Babylonian astronomers and through them is embedded in today's measurement of time: the "1:12:33" on a computer clock means 1 (x 60-squared) second + 12 (x 60) seconds + 33 seconds...
https://www.nytimes.com/2010/11/23/science/23babylon.html

Fascinating, isn't it, that we have assimilated their sexagesimal system and still use it today in timekeeping! Next time you glance at your watch think of those ancient civilizations. Akkadian and Sumerian tablets predated Babylonian and Persian, but the number system remained the same even after the tower of Babel fell and God confounded their languages. The numbers remained, but their alphabets morphed and scattered to new regions and nations. Please allow me to digress here for a minute... fyi, from there (old Babylon) it is believed by many that Nimrod, the son of Cush and the great grandson of Noah and the King of Shinar, retreated south with others who spoke the same (new) language as he did, and was the first pharaoh of Egypt, Narmer (founder of the first dynasty). Some have even proposed that Nimrod and Enmerkar, a Sumerian King of Uruk/Erech (which is probably where modern day Iraq got its name) are one and the same. The descriptions of these different archetypes seem to point to Nimrod. I wouldn't doubt any of these propositions because, after all, the languages did change, and the phonetics would have changed too. In fact, it is becoming more clear by modern researchers, (ie, Dr. Tom Horn, Rob Skiba, Trey Smith, Alexander Lawrence, to name a few) that many ancient pagan gods in various locations around the world can be traced back to Babylon and originate from the same "god." It makes simple sense that pagan gods can be traced back to Mesopotamia. To Babylon. Nimrod, who built the tower of Babel, is believed by many today, as one of the giants from the old testament who was worshiped as a god. We get the word babel from this Bible story. Babel means noise and confusion.

Moving on… forgive me for the irrelevant tangent. I couldn't help myself… although the part about Babel, is somewhat relevant since we are talking about various numerical glyphs.

I	II	III	IIII	Γ	ΓI	ΓII	ΓIII	ΓIIII	Δ
1	2	3	4	5	6	7	8	9	10

ΔΓ	ΔΔ	Ⅎ	H	ℍ	X	⋈	M	ℳ	)
15	20	50	100	500	1,000	5,000	10,000	50,000	¼

(	⊢	ℿ	Δ	Ⅎ	H	ℍ	X	⋈	Ⅎ
½	one drachma	five talents	ten talents	50 talents	100 talents	500 talents	1,000 talents	5,000 talents	five staters

Δ	Ⅎ	H	ℍ	X	M	ℳ	A
ten staters	50 staters	100 staters	500 staters	1,000 staters	10,000 staters	50,000 staters	ten minas

Figure 20. Ancient Greek (Table of "Attic" numerals)

Greece eventually changed from Attic to Acro-phonic numerals using their alphabet system. Compare it to the Hebrew numeral/alphabet system (figure 21). It appears to me that the Greeks may have copied, or rather, were inspired by the Israelites. Curiously similar, wouldn't you agree? The alphabet too. The Hebrews were original in their numbers, having derived from their alphabet, they did not separate the system of letters from numbers which gave/give deeper meaning to the numbers and significance in Bible prophecy, and in Bible micro-codes and macro-codes. It is also noteworthy to mention that the Hebrew calendar is based on Anno Mundi which translates from Latin as *in the year of the world.* The Hebrew calendar is much different than the calendar we all use today, but its lunar months and holidays are still very much observed by Judaism and Messianic Jews alike. According to the Encyclopedia Britannica, Rabbi's calculated elapsed time from the accounts in Genesis and added the genealogical record in Genesis to reconcile their widely accepted date count. The year is 2020 according to the Gregorian calendar, but it is 5780 according to the Hebrew calendar.

THE BIBLE - ALPHABETS AND NUMERICAL VALUES

		Hebrew					Syr			Greek			
OV		ARL	TNR	NV	P	Name			ARL	TNR	NV	P	Name
1	𝕏	א	א	1	(a)	Alef	≺	⅄	A α	A α	1	A	Alpha
2	𝟡	ב	ב	2	B, V	Bet	⊃	B	B β	B β	2	B, V	Beta
3	٦	ג	ג	3	G	Gimel	↘	Γ	Γ γ	Γ γ	3	G	Gamma
4	△	ד	ד	4	D	Dalet	ᴦ	Δ	Δ δ	Δ δ	4	D	Delta
5	⅁	ה	ה	5	H	He	௰	Ε	E ε	E ε	5	E	Epsilon
6	Y	ו	ו	6	W, V	Vav	๐		F, Ϛ	F, Ϛ	6	W	Digamma*
7	I	ז	ז	7	Z	Zayin	؞	Z	Z ζ	Z ζ	7	Z	Zeta
8	月	ח	ח	8	H, X	Chet	↙	H	H η	H η	8	E	Eta
9	⊗	ט	ט	9	T	Tet	↳	Θ	Θ θ	Θ θ	9	Th	Theta
10	?	י	י	10	J, I, Y	Yod	؍	I	I ι	I ι	10	I, J	Iota
11	ℐ	כ ך	כ ך	20	K, X	Kaph	↖	K	K κ	K κ	20	K, C	Kappa
12	6	ל	ל	30	L	Lamed	⅃	Λ	Λ λ	Λ λ	30	L	Lambda
13	ɱ	מ ם	מ ם	40	M	Mem	҂	M	M μ	M μ	40	M	Mu
14	ℓ	נ ן	נ ן	50	N	Nun	‿	N	N ν	N ν	50	N	Nu
15	丰	ס	ס	60	S	Samekh	∞	Ξ	Ξ ξ	Ξ ξ	60	Ks	Xi
16	O	ע	ע	70	(o)	Ayin	↲	O	O o	O o	70	O	Omicron
17	⌐	פ ף	פ ף	80	P, F	Pe	؎	Π	Π π	Π π	80	P	Pi
18	Ͱ	צ ץ	צ ץ	90	Ts	Tsade	ؼ		Q, ͱ	Q, ͱ	90	K, Q	Qoppa
19	Ϙ	ק	ק	100	K, Q	Qoph	؋	P	P ρ	P ρ	100	R	Rho
20	⊿	ר	ר	200	R	Resh	؉	C	Σ σ,ς	Σ σ,ς	200	S	Sigma
21	W	ש	ש	300	S, Sh	Shin	؊	T	T τ	T τ	300	T	Tau
22	✗	ת	ת	400	T	Tav	⅄	Y	Y υ	Y υ	400	U, Y	Upsilon
23								Φ	Φ φ	Φ φ	500	Ph	Phi
24								X	X χ	X χ	600	Kh	Chi
25								Ψ	Ψ ψ	Ψ ψ	700	Ps	Psi
26								Ω	Ω ω	Ω ω	800	O	Omega
27									T, ꓤ	T, ꓤ	900	Ts	Sampi

The ancient Hebrew alphabet has 22 letters (from Alef to Tav). The earliest characters originally consisted of symbols (first column), from which the today's digital signs originated. Five of them have special end forms (e.g., Kaph). The letters Alef (1) and Ayin (16) have a sound that does not occur in our language, so their pronunciation is in brackets. Although the Syriac (Syr; Aramaic) language is similar, it has produced completely new letters over the centuries (see Peshitta).
The ancient Greek alphabet has 27 letters. In addition to the 24 main letters (from Alpha to Omega) in which the NT was written, there were 3 more characters in the early form of the Greek alphabet; i.e. in the alphabetic ("Milesian") system of Greek numerals: Digamma* (F, ϝ; [w]); since the Middle Ages Stigma in the appearance of a snake: ς, ϛ; NV 6), Qoppa (NV 90) and Sampi (NV 900). Both alphabets have together **22+27=49** (7x7) letters. The whole Bible (OT and NT) was written on **22+27=49** (7x7) scrolls.
Abbreviations: (OV): Ordinal value, the order of the letters in the alphabet. (NV): numerical value (Gematria = Interpretation of words by numbers). (P): Pronunciation or Transcription. The letters may look different in some sources because they are often shown in multiple fonts, e.g. Arial (ARL) or Times New Roman (TNR). In Greek, there are uppercase or lowercase letters.

Figure 21. Ancient Greek vs Hebrew Numerals.
Notice the many similarities

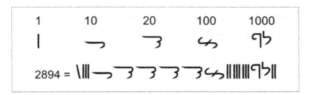

Figure 22. Aramaic Numeral Signs

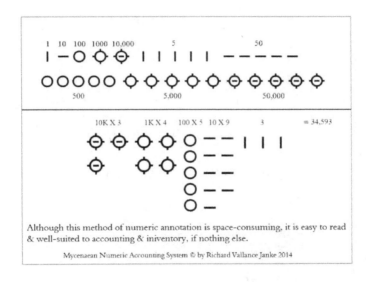

Figure 23. Earliest Greek syllabic script, Mycenaean
Linear B Numeric System

Counting became increasingly more complex, I'm sure, when bartering in marketplaces demanded fair trade. Various nations had symbols in their own number systems. For them to account for higher magnitudes of value a symbol was represented. It was still cumbersome, I suspect, especially when dealing with bulk purchases. Lol! The abacus came in handy, no doubt! Though the exact origin of the "counting frame" is a mystery, it was used in Ancient East and Far East, as well as Russia and Europe as a rudimentary calculator long before the Hindu-Arabic numeral system.

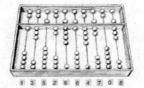

Figure 24. Abacus

中文數字大寫對照表													
0	1	2	3	4	5	6	7	8	9	10	100	1.000	10.000
〇	一	二	三	四	五	六	七	八	九	十	百	千	萬
零	壹	貳	叄	肆	伍	陸	柒	捌	玖	拾	佰	仟	

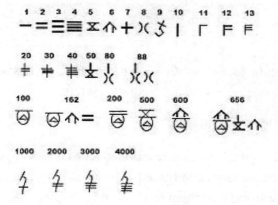

Figure 25. Modern Chinese Numerals

Figure 26. Chinese Shang Numerals, circa 1400 BC

Notice the similarities to their modern glyphs. You would think their numerals would have become simpler, yet the contemporary glyphs are more complicated.

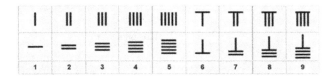

Figure 27. Rod Numerals used by Chinese merchants,
Han Dynasty to the 16th Century

Figure 28. Chinese Suzhou numerals used in bookkeeping until the 1990's

China has many more numerals on record not shown here. Perhaps the Chinese take the cake for having the most interesting, diverse, and complicated history of the numerical system. Or the Mayans! And the winner is...

The Chinese, as well as the Babylonians, and even the Aztec/ Mayans were known to use a positional value system in their counting sequence to make it easier to notate higher numbers. You can see from the charts that the Babylonians and Mayans had a zero in their numeral systems, but by looking at the Babylonian system it would be hard to identify because their "empty (zero) place holder looked more like a chevron sideways than a dot or a circle or an eye or pie. Lol! Nonetheless, it represented the zero. A place holder between digits for higher values would rid the confusion of 60 as opposed to 600 or 67 as opposed to 6007.

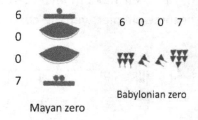

Figure 29. Mayan vs Babylonian zero

According to the history of numbers, 8th century India perfected this method, so the trophy goes to them! In fact, the decimal system began with them also. It was an Indian astronomer and mathematician, Brahmagupta (I'm going to call him Brahm, for short), in 628 AD, who first introduced the idea of using a dot (as a zero) for mathematics, and it was Aryabhata (I'm going to call him Arya), another Indian mathematician, who proposed the zero (a dot) as a place holder for notations. This became simple and useful when denoting higher values. Instead of placing a symbol for each number group for large numbers we now had zero (a dot) to denote the higher value! This concept also made calculations consistent and reliable. Can you imagine solving calculations with picture symbols? It was all coming together. We (humankind) had a zero, finally, as the place holder for the base power of 10. Woohoo! Brahm, while computing equations using the dot, started writing them down as open circles, thus was born the zero as we know it today, thereby giving birth to mathematics as we know it today.

For the record, there are other civilizations who had advanced number systems which are not mentioned here, but I couldn't include everybody. Sorry.

The other integral contribution to arithmetic came from a Persian Arab from the 9th century, named Al-Khowarizmi, who wrote Al-jabr (The Compendious Book on Calculation by Completion and Balancing). He was the father of algebra which is considered the foundation of math. Together, the findings with al-Kindi (another Arabic mathematician) and the numeral system given to us from Brahmgupta and Aryabhata (oops, I mean Brahm and Arya) merged to form the Hindu-Arabic or Arabi-Indi system. It was from here that it evolved into what Europe would eventually adopt... and institute the numbers we use today, established globally. All from the Hindu-Arabic base-10 decimal number system.

Hindu-Arabic
NUMERAL SYSTEM

○	0
?	1
২	2
३	3
४	4
५	5
६	6
७	7
८	8
९	9

30. Hindu-Arabic... old to modern

Indian numerals had a variety and a history of evolving symbols much like the Chinese. Different regions giving birth to different glyphs.

31. Sanskrit Arabic Numerals

32. Brahmi Numerals

Zero is an amazing number. Correction: non-number. Because it is a non-number it was not well received by the western world. The Romans rejected it and in some countries it was outlawed. It wasn't until Leonardo Pisano Bigollo, better known as Fibonacci, used it in his formulas when Europe finally welcomed zero to its numerical systems. In chapter nine we will learn more about this Italian mathematician and his fascinating "sequence" discovery!

Now that we've established the birth of zero we can return to the paradox of time keeping. As you know math is essential for keeping time in sync with nature. It was especially important to have a calendar congruous to seasons of planting and harvesting. This was evident in ancient times, as well as modern history. Our current calendar, the Gregorian Calendar, was made nearly perfect after integrating astronomical events to the time-tracking table. No more sundials or water-clocks! No more mistakes, right? Today we can rely on precision timing, the NIST-F1 Cesium Fountain Atomic Clock at the National Institute of Standards in Boulder, CO is as precise as they come. We can relax; they track time using "hyperfine transition frequency in the microwave, or electron transition frequency in the optical or ultraviolet region of the electromagnetic spectrum of atoms." See https://en.wikipedia.org/wiki/Atomic_clock. Farmers today know exactly when to sow their seed and when to reap their harvest. The church can rest easy too. No more issues on when to celebrate Easter!

The purpose of this chapter is not to examine absolute zero which, by the way, is zero on the Kelvin scale and equal to -273.15°C or -459.67°F. That has to do with temperature, not time, right? Or does it? Isn't it possible that the global warming we are experiencing in our "common era" has to do with the cosmic celestial season we're in, and the timepiece of our solar system has reached an hour where things are really heating up? (Fun pun, lol... I know I am ridiculous). If so, will minimizing our carbon footprint really make a difference in climate change? In the cosmic scheme of things? *Sorry for the tangent, but I really think the above statement should not be overlooked. And I wouldn't be surprised if we will have to adjust our clocks and calendars again in the not-too-distant future. Perhaps our maps as well, considering the pole drifts of our earth's axis according to variations of the cataclysmic pole shift hypothesis.

Okay, out of that rabbit hole and back on track...

Is it a coincidence that zero is a circle? What is a circle? What if Brahm and Arya didn't use a dot which then transformed into a circle? What if they used the chevron symbol of the Babylonians? Or the eye, half-pie looking thing of the Aztecs? Do symbols matter? And is it any coincidence that the circle was used to denote the concept of zero, zilch, nada? At the same time, it represents perpetuity, right? Isn't that why, traditionally speaking, marriages are symbolized with a ring? A circle? *I'll love you forever... til death do we part.* While zero is the culmination of absolutely nothing it's also the consummation of continuity. Coincidence?

The concept of zero was understood as a place holder between sets or groups of numbers. It didn't count for anything in of itself, therefore, had a value of none. A void. Empty space. Yet, it had a remarkable purpose. To enter into the next level. Like a gate. Without it we wouldn't be able to multiply like we do. In fact, without zero, we wouldn't have algebra or calculus. We wouldn't have computers. Technology would not be where it is today without zero. According to Moore's Law the number of transistors in an integrated circuit doubles every two years. Can you imagine operational algorithms without zero!

Zero is nothing yet everything simultaneously. It's not only a place holder and a gate, but it seems to serve as an anchor, as well. Take the Cartesian Coordinates, for example:

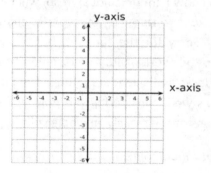

33. Cartesian Coordinates

It is more commonly known as the X Y Axis. Just look at the image (figure 33). There are two perpendicular planes (Cartesian planes), and zero rests at the center of the axes. An explanation is not necessary. It appears that zero serves as an anchor for all formulas pertaining to it. Geometry and trigonometry would be lost without it! It is quite literally central to the system. Agreed?

In the natural order of things, we don't start with zero, we start with one. But we end with zero and begin again with one... yea, technically, we end with 10 (or 20, 30, etc.) and it counts in the counting, but it is more than just a number. It's a gate. Every time we come back to zero another generation is born. If it weren't for 10, I don't think we ever would have realized we needed a zero between 1 and -1, and in this case it serves as a gate going backwards. But 10, 20, 30 and so on serves as gates going forward. Once you reach ten you come to zero again; the only reason we put a one in front of it (10) is to signify we are entering into the 2nd gate. The 2nd generation. Much like the paradox of time, even though it is a one, it is the 2nd order. It is the 2nd set of base 10. 20 is the 3rd set, 30 is the fourth, etc. This can go on forever; 300, 3,000, and all the way to googol (1.0×10^{100}) and beyond!

Look down at your hands and count your fingers. Unless you are an alien with four fingers (or a giant with six, lol) or have Polydactyly Syndrome or conversely if you are short one or few fingers (or hand(s) or arm(s) for that matter) from an accident or birth-defect, I'm sorry, truly. Forgive me, this will not pertain to you. Look down and count. Know this: your tenth finger is the gate. The zero. Ten represents a generation of the whole set before it. It counts in the number sequence as it functions as an end and start point. You may have ten digits, five on each hand, but there are only 9 *individual digits* in math. Both zero and 10 are gates between them. It is a place holder. It is a gate. It is a puncture allowing a new set of digits to come through and begin a new generation. It is a ring of perpetuity and a hole of a birth canal. The Rodin Coil (more on this in chapter nine) using the torus energy model which moves in perpetuity dismisses the number ten in its system. It uses only 1-9. Why?

It's paradoxical because zero counts in the count when counting units, but it is mysteriously unnecessary in perpetual motion as seen in vortex math. A vortex is a "rapid rotary movement of cosmic matter about a center, regarded as accounting for the origin or phenomena of bodies or systems of bodies in space (in Cartesian philosophy)." See dictionary.com.

Why does a vortex work without the number 10? Here, the axis point moves in three dimensions. But zero tethers all dimensions, not just three.

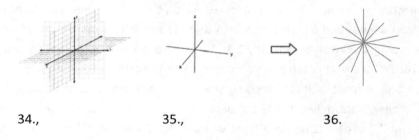

34.,   35.,   36.

Is zero multi-dimensional? Does it go beyond space and time? Is zero inter-dimensional as well as self-contained and self-perpetuating at the same time? Is it like a black hole or a worm hole (Einstein-Rosen Bridge) in the fabric of space? Consider the counting numbers; the sets don't just move forward. They move backward. The multi-dimensional star you see above from the cartesian coordinates (figure 36) doesn't have lines moving outward only from the center axis. They have lines you don't see moving inward, as well.

Zero, absolutely, is a place holder and a gate. It appears to be an anchor and a vortex too, but these assertions call for an atomic knife. We must unravel the circle.

Peak Priority

Scrabbling up a scree field
only to smear yourself
against a wall of granite.
Scramble a score of boulders.
Fold to fit in a fissure
a crevice creeping to the sky.
Cracks made crooked by a knife unknown.
A quakes' carving or
maybe waters' way.
It appears to you
wedged in a wall well above the western tract.
Feet gripping.
Grazing palms agree
grasping granite stings as it scathes.
Mantling exhaustively
to cross over a crux of a corner
and creep over the peak
to find the mount
underneath your feet
and a twilight greeting from Draco the Dragon.

1997

Chapter SƐVƐN

Mile and ½ High

It was the same rock face we had climbed before. The thrill of it lost its potency and now we were searching for a new route to climb on that same rock face. The backside of Hartman's Rocks was less crowded and was easier to access, in that there were no hills to climb, or winding and fitful, deep divot jeep trails to cross to get there. The backside, however, was farther away from town, on the way to Blue Mesa Reservoir. There were more shrubs and trees than on the front side. All the green could probably be attributed to the fact that we were nearer to the river in proximity and elevation.

Hartman's was several miles southwest of Gunnison. It was (and still is) a mecca for mountain bikers, rock climbers, off-road vehicles, and mountain runners. Hartman's Rocks Recreational Area is to Gunnison as (world-renowned) Slick Rock Rec Area is to Moab, Utah, but not quite on as large a scale. Yet, it had miles and miles of jeep trails and single tracks running and twisting along hilly-rocky desert terrain spotted with sagebrush, evergreen, cacti, and flowers (many wildflower blooms in the

Spring). It was surrounded by granite outcroppings mainly at its eastern end, some towering as high as 75-100 feet, or maybe higher. It was perfect for novice climbers learning the sport, like me. Plenty of cracks and sturdy knobs to hang onto, fortified by igneous slab rock. It was also ideal for those without climbing gear to boulder freestyle among the smaller slabs. I was totally familiar with the front side of Hartman's, having biked there all the time. The back side was becoming more familiar as we chose to climb there on a semi-regular basis, lately.

It was a beautiful area! More shady. More private. The perfect place to pull out the pipe, pack it with P-Bud, and pass it openly to our small group of thrill seekers as we readied ourselves. "P-Bud" was the name given to the most desirable marijuana you could find in the area. If it wasn't P-Bud it was probably *Maui Wowee*. You just couldn't find good local pot unless it came from Paonia, and if it didn't come from Paonia or Hawaii, it was just weed from God-knows where. No one wanted weed. *"Kind bud" is where it's at!* Sticky flowers of Cannabis, so dense and potent you could smell it a mile away just like Colorado's Western Spotted Skunk (Spilogale Gracilis). Lol!

Paonia, Colorado was not too far from Gunnison, as the crow flies, but driving there was quite a trek. You would either have to continue on County Road 50, and cut right to 92 through to the Black Canyon on the far west end of Blue Mesa and go north a long ways until you reach the 133 and go northwest from there… about 83 miles total. The other way, somewhat of a shortcut, is using Kebler Pass via Crested Butte. Unpaved. And as you recall from the chapter detailing my snowmobile accident, Kebler Pass is only open to automobiles in the summer months. P-Bud was not easy to come by, but coveted year round and when a batch came to town, you had to be at the right place at the right time to get some or it would be gone until the next batch arrived. During the dry spells, weed would have to do.

Not today! One toke of the pipe full of P-Bud was all it took to get you to that place higher than what we planned to climb. Not far from our usual route (climbing route on the rock) was a smaller outcropping tucked behind some trees and a bit farther up an incline. It was my first time on this route and as I scouted out the line I wanted to take I realized it would be the most challenging climb to date, at least for me, the newbie! My boyfriend, Caleb, on the other hand, was no neophyte; and he taught me everything I knew in regard to climbing and what I'm about to share with you now. Much has changed in the world of climbing since the early 90's, so forgive me if I'm not up to par.

A quick explanation of sport climbing most people do not know: Sport Climbs are typically pre-routed... bolts are drilled in the rock and in place already or, today, handholds on artificial walls, like you find at REI or on the back of a Royal Caribbean Cruise Ship (the one and only time I ever climbed a fake wall). Whereas, Traditional Climbing (which we did) takes place on natural rock (outdoors) and requires the climber to bring their own tools (carabiners, hexes, cams, nuts, etc. hanging from what you call a RACK), which is to assist in the climb and used to set anchors for protection as you climb. Traditional Climbing (aka trad) should be done with caution and with someone who has expertise in the sport. There are other types of climbing, such as bouldering and cragging, which I will not get into... and big wall climbing, which was what my boyfriend, Caleb, grew up doing. Big wall climbing is exactly that... BIG WALL climbing, like climbing the famous El Capitan at Yosemite National Park in California. Or Devils Tower National Monument in Wyoming. Climbing big walls thousands of feet high required multiple pitches which means it is more than a ropes length high. What we did at Hartman's Rocks was single pitch climbs, no more than 100 feet max... it required one rope's length...

Caleb grew up climbing big walls with his Green Beret uncle in the Smoky Mountains of North Carolina. Climbing a big wall takes much preparation and a lot more gear to carry... up the cliff. In their case, they would bring cliff beds, sleeping bags, and tents, fleece parkas, food and water, and extra gear on full racks because they would spend the night hanging from a rope on the wall. Sounds scary I know, but Caleb assured me that they took every precaution and never rested without, at least, three anchors set in the rock above. Climbing Big Walls required more rope as well, after all, you are climbing multiple pitches (about a thousand feet a day), and you have to haul up the bag carrying your camping gear!

As for us college students looking for a high no more than 70 feet, and stoned harder than the rock (lol) we were set with one rope and a half rack. We had just enough tools on the rack for Caleb to lead the climb, get to the top, set anchor, and repel down with a "top rope" in place. That's what I did. I was a "top-rope" climber which meant I was secured to an anchor, three of them. Although we didn't have the same payload as Caleb and his uncle would have on their big rock climbs, Caleb always erred on the side of caution and insisted on setting three anchors at the top. I was grateful. As the "lead" climber, Caleb would go first, carefully selecting the line or the route he knew I would be comfortable with... on this day a 5.9., except for the bottom which appeared to be more of a 5.10 or 5.11 due to the cave at the bottom of the cliff.

Climbs are classified by difficulty. 5.0 to 5.7 is considered to be an easy climb and anything over a 5.11 is considered hard. 5.13 to 5.15 is for experts. I was a beginner, but a fast learner. This was my 2nd season climbing and I guess I was more than a novice by now due to the number of climbs I had under my belt. I don't know exactly how many climbs you have to have made before graduating to an intermediate class, but I know by this time I was climbing 5.9 to 5.10 regularly on granite. It is much more different when climbing on, say, sandstone, like we did at Potash in Moab, Utah. For the record, I do not like sandstone, especially sandstone crack climbs. It's hard on the hands... both sides of the hands! You wedge your hand in the crack and then turn it or make a fist, whatever it takes, for your hand to be used as the tool, the nut (if you can imagine) to hold you as you make your next move and do the same with your other hand and so on until you reach the top. Sometimes my hand would get stuck in the wedge of the crack because I would plant it so deep for fear of falling, and then it would cause me to back climb down in order for me to release my hold and place it in a more forgiving position so I can then release it with ease when I moved above it. The memory of it alone makes me want to go put O'Keefe's Moisturizer on right now!

Granite's where it's at! I mean it. Don't get me wrong, it can still scathe. I mean rock is rock, right, but, if you are considering climbing in nature I recommend you stay away from sandstone. Choose granite. It's easier to grasp and easier on the skin.

This climb ahead of us had its crux at the bottom. The "crux" is the hardest part of the climb. Usually, on the routes we climbed regularly, the crux was at or near the top or middle of the climb. This one was different and the hardest one yet. I stared at the crux as I put on my harness, deliberating to myself what would be my first move. This cave at the bottom was pretty deep so we would have to climb around it. The line (or route) we would take beyond that was straight above the middle and deepest part of the cave, and the climb would get easier from there. Even if we stuck to the sides we would be somewhat inverted for the first couple moves. I was nervous but glad these moves were not far off the ground. The cave was maybe 15-20 feet wide, and about as high as it was wide. It was deep enough to take shelter and to stay dry from a hard downpour of rain. Not a true cave by definition, but more than just a concave in the cliff.

Today was a beautiful day. Sunny and warm. We were all in shorts and t-shirts. We each chimed in and gave our opinion on how to get around the crux, as we passed the pipe and put on our climbing shoes.

Caleb, who was already ready to go came over to attach his rope to me. I finished tightening my laces and stood. He took the colorful rope, which was already attached to him, pinched, and looped it into the belay device and a carabiner which he then locked to my harness. Whoever's responsible for the belay device is subsequently called "The Belay." As the belay I was responsible for giving slack or taking up the slack of the rope as Caleb made his ascent. Caleb, being the lead, would climb without the protection of a top rope anchor. He would set temporary anchors on his route up, approximately every 10-20 feet or so. If he were to fall before he set the first anchor he would hit the ground, but the fall would only be approximately 10-20 feet. Once he sets his first anchor, he would loop his rope into a locking carabiner attached to the anchor, securing him from falling to the ground. Now, if he were to fall, he would only fall the distance from the anchor times two. For example, let's say he sets an anchor and ascends another 5 feet before setting the 2nd anchor and slips and falls, he will fall the five feet he had just climbed (from the first anchor) and another five feet, the slack of the rope before it meets the 1st anchor. From this equation he will fall 10 feet.

There is another variable, however. Me! Since I am the belay it is my responsibility to keep my eyes and ears open to the climber, especially if the climber is The Lead. Unlike me, Caleb liked a lot of slack when he climbed. If he could feel the tug of the rope he would holler, "Slack!" and I would give it to him. You see, at the beginning of the climb, most of the rope is on the ground, carefully laid in a circular pile, so as not to tangle. If you followed the rope, it would come to me, through my belay device, and onto Caleb, securely attached to his harness in a figure eight knot. As he moved away/upward there would be more rope between us. I had to make sure both my hands were on the rope; one on the top of the belay device moving toward Caleb and the other from the slack on the ground. If both my hands were gripped firmly on the rope from both sides of the belay device and in its appropriate position, the rope would be locked and Caleb would not be able to move, but as I moved the rope and loosened the position and grip the rope would be free to move through the belay. In fact, we used belay devices to rappel, as well.

This day we had two ropes. One for the climb, and the other to anchor me to a tree! Yup you heard me right. I was anchored to a tree! That's because the day before while we were climbing our usual route, Caleb took a tumble. That's what you call it when you fall in climbing. That's what he called it, anyway.

"I learned to tumble when I was a kid," Caleb would say, "but, you never get used to it." There's a right way and a wrong way to tumble, much like what they teach you in Gymnastics. You want to roll with it to lessen the impact on your body.

Before his tumble, Caleb called out for more slack, and I gave it to him. Boy, did I ever. Probably more than I should have, but I took for granted his ability, and the fact that he had never fallen... with me. To make matters worse, my hands were off the rope when he slipped. At least one hand was... it happened so fast. One minute he's climbing making headway, setting anchors, halfway up the route. 40 or 50 feet off the ground. Next minute he's falling, and my rope is burning through the belay device and my hand. Thank God for gloves! He had several anchors set which slowed the fall. Barely. 180 lbs and gravity are hard to contend with and it literally lifted me off the ground once I grabbed hold of the rope with both hands and positioned them to lock. He would have hit bottom in a near free-fall had it not been for my quick response and counterweight. He fell nearly 15-20 feet and came away with only a few scratches thanks to his ability to tumble. Had both my hands been on the rope, and the slack taken in he would have only fell approximately ~~five~~ ten feet because he had just set an anchor beforehand. And I would have remained on the ground, we think. But we didn't want to take any chances today, after all, he was 180 lbs and I was 130lbs.

Caleb's fall amounted to five feet times two plus the amount of slack I gave him plus the amount of rope that was able to escape from my belay device which was not in a locking position. 5x2+ the simple formula of distance traveled.

Distance = time/ velocity. Since we didn't have exact measurements I am going to hypothesize the numbers we do "know." If he fell five feet (the distance from the last anchor) x 2, plus the extra distance caused by slack, another 10 feet, we can finish this equation with these approximate numbers.

(5x2) + 10 = t/v or 10 + 10 = t/v or 20 = t/v

Since we don't know how long it took for Caleb to fall 20 feet, we will refer to wiki for a usable number for velocity:

> Near the surface of the Earth, an object in free fall in a vacuum will accelerate at approximately 9.8 m/s^2, independent of its mass. With air resistance acting on an object that has been dropped, the object will eventually reach a terminal velocity, which is around 53 m/s (195 km/h or 122 mph) for a human skydiver.
> https://en.wikipedia.org/wiki/Velocity

Caleb was very near the surface of the Earth, so we will use 9.8 m/s^2 (meters per second squared) for velocity.

20 = t /(9.8m/s^2)

...but wait, there's another factor. Caleb was not in a "free fall." He had resistance (r) to help break his fall because he was tied to a rope woven through metal carabiners. Although the friction the rope and carabiners would have caused were minimal they were another variable, nonetheless, which we must take into account.

$20 = t/(9.8m/s^2)-r$

Here r stands for the resistance, so now we must find a formula for "drag." Do you see how simple equations get more complicated? If there is something that will affect it, it must be taken into account. In this case the "drag" or resistance which slowed down the speed of Caleb's fall. Then we must consider the many possible factors to determine what caused the drag. The friction of the rope moving through carabiners, or the rocks' face, etc. There are many variables, and even if we use the standard formula for drag, the solution will not be 100% accurate. According to Sciencing.com The Drag Force Equation varies depending on the "force per unit area" which, essentially, is the pressure proportionate to the object. For your pleasure, I will allow them to explain, lol (skip it, *if you must*):

> Drag forces, being restrictive rather than propulsive, are not as dramatic as other natural forces, but they are critical in mechanical engineering and related disciplines. Thanks to the efforts of mathematically minded scientists, it is possible to not only identify drag forces in nature but also to calculate their numerical values in a variety of everyday situations...
>
> **The Drag Force Equation**
>
> Pressure, in physics, is defined as force per unit area: **P = F/A**. Using "D" to represent drag force specifically, this equation can be rearranged to **D = CPA**, where C is a constant of proportionality that varies from object to object...
>
> **Drag Force on a Falling Object**
>
> One of the equations of motion for an object in free fall from classical mechanics is **v = v₀ + at**. In it, v = velocity at

time t, v_0 is initial velocity (usually zero), a is acceleration due to gravity (9.8 m/s^2 on Earth), and t is elapsed time in seconds. It is plain at a glance that an object dropped from a great height would fall at ever-increasing speed if this equation were strictly true, but it is not because it neglects drag force.

When the sum of the forces acting on an object is zero, it is no longer accelerating, although it may be moving at a high, constant speed. Thus, a skydiver attains her terminal velocity when drag force equals the force of gravity. She can manipulate this through her body posture, which affects A in the drag equation. Terminal velocity is around 120 miles per hour.
https://sciencing.com/how-to-calculate-drag-force-13710258.html

As you can see, there are many variables, and to get a true number for an answer would be nearly impossible. The quotes above are just an abbreviated explanation of the drag force equation and in free fall they refer to atmospheric drag and not mechanical drag. I did not even include their equation on fluid drag (as with a swimmer). It's enough already to make our heads spin, right? We would end up with an approximation, at best. In the case of Caleb, the drag was not caused by atmosphere, yet there may have been some aerodynamic drag in addition to the rope resistance. It just seems to get ever more complicated... and I don't know about you, but I'm ready to put this to rest. The point here is not to determine how fast Caleb was falling. Who cares! I just want to make clear the impossibility of knowing the exact answer due to the unknowns. In the grand scheme of things this is a very simple equation compared to so much else we could attempt to measure in the world. This scenario is one small example. Formulas would seemingly get bigger with more complicated scenarios, right? The solution is beside the point.

Let's continue the story so I don't leave you hanging… hehe, lol!

"I think you were lifted up further than he fell," Caleb's friend chuckled as he helped to tighten the rope to the tree, remembering the day before when Caleb's fall caused me to fly off the ground.

"Farther," I corrected.

"What?"

"Farther. You said further. It's farther. When referring to physical distance the correct word is farther." I continued. I couldn't help myself. I used to be insufferable that way sometimes.

"Oh ok," he rolled his eyes. "You sure you don't want me to belay?"

"No, I got this," I replied flashing a knowing yet apologetic look to Caleb, as he glanced over after securing his rack to his harness. It amazed me that Caleb was going to do it again after such a traumatic day before. If it were me, I would have given up the sport altogether. Falling is one of the most frightening things a human can experience, even if you are tied to a rope. Or a parachute. Or a bungee cord.

"Climbing!" He yelled as Caleb took hold of the granite before him.

"On belay," I bellowed back to let him know he was secure.

He rounded the mouth of the cave with ease. Not much of a crux to him. He demonstrated an inversion technique to us watching, so we could copy him when we had our turn. He took his time and explained why he did what he did until he got above the cave, at which point he streamlined the rest of the climb swiftly. Perfectly. Once he reached the top my job as belay was over. He would not need my assistance to lower him down.

After he set three anchors, he self-rappelled down stopping at each temporary anchor (he had set on his way up) to remove them. They were no longer necessary now that the rope was anchored securely in place x 3 at the top of the climb.

It was my turn. Caleb and I basically switched places. He untied the knot from his harness and tied it to mine. He took the belay device from mine and secured it to his. I didn't need his rack, however, because tools were no longer needed, he took care of that! I was "top-roping." There was nothing to fear. If I slipped, I would barely fall an inch because he always kept me on a tight rope. In stark contrast to him, I would always holler, "tighter" when I didn't feel the tug of the rope on my harness. If it weren't for that tugging feeling I may have never attempted to climb and got off the ground in the first place, giving into my fear of heights. The tug would sometimes pull me up without having to move a muscle. My muscle, not his, lol!

The top rope was set dead center of the cave, and to get around the cave you had to go either to the right or left perimeter of the cave causing the rope to veer off center. If I slipped, I would move more than an inch this time. I would swing to center. Could be fun, I thought, as I took a hit from the pipe, held it in for two seconds and coughed as I exhaled. Caleb's route was to the right of the cave so there I went, attempting to copy his move. My right hand gripped on a good hold, my left searching for an equally good hold. Found it!

"Climbing," I hollered, one foot already off the ground, pushed in a little pocket in the rock.

"On belay," he replied, standing not but five feet behind me.

I found another pocket for my other foot, and I was officially off the ground! *Wow, I'm inverted,* I told myself. Wish I had a picture of this! The next move I made felt okay but I slipped and went swinging to the left of center and back again. Weeeee! Six inches from the ground. I would have dragged my feet if I hadn't tucked in my legs!

The 2nd and 3rd time around I decided to traverse farther to the right of the cave, way off center, but not as inverted. A trade off because there weren't as many pockets, cracks, or knobs to hang onto. I felt it was worth the try. As one move after another took me round the top of the cave I realized I made it. I got passed the crux. Whew! I felt on top of the world, but I was no more than 20 feet up. There was a good three-inch ledge to rest on and so I did. I was already a little winded. I caught my breath and saw that I was centered and now moving straight up. There were plenty of knobs and crevices to hold onto. They were calling out my name, so I started to climb again without calling it. It is considered climbing etiquette to yell "climbing" each time you begin to climb even after a 10 second break, to give the belay a... heads up... hehe!

This was probably the one time I had not noticed the "tug" of the rope. I must have had an extra dose of confidence with that last dose of THC. The rest of the climb appeared to be a breeze compared to the crux! With my next move I lost my footing and I couldn't catch myself in time. I was falling! I screamed as I tumbled. Your first instinct is to grab the rope, but that's no use because the rope is falling with you. I reached out to grab the rock face only to find it disappear above me as I swung into the cave abruptly. Startled yet relieved that a pleasant swing was at the end of my fall. I fell about eight feet. Doesn't sound like much, but it seemed like a long fall to me! Caleb said that was a good fall.

Enough to scare the daylights outa ya!

"Karma's a b**tch," he laughed, and then apologized. Then I apologized.

"It was my fault, "I said, still swinging back and forth, "I didn't let you know I was climbing. I just assumed you were watching. I thought you knew." Caleb lowered me slowly to the ground. I didn't have it in me for another go of it. The sudden adrenaline took a lot out of me and did me in! I was done. But I didn't give up the sport, after all. I was back the next day and conquered the climb!

*I wish I had the same fearless fortitude to climb *God's mountain* back then!

Rope's End

Each crease in my cheeks has turned from crevice to crevasse
The spots on my hands to continents,
and incontinence has me dependent on Depends.
My friends have lost me in their memory,
and my primary goal each day when I wake is to ease the pain
in my joints, breast, and spine.
The shine of the sun is as a hundred bees stinging my face
as I race across the courtyard in my chair.
Farewell am I when the girl holds my hand
as she places a spoonful of creamed rice in my mouth.
To the south I travel once a month
to visit the grave of my one and only son.
To run once more on a sandy beach is my dream each night while I sleep,
and to leap into the light once and for all
is my dream each day while I weep

1999

Chapter Σ1GHT

Ick

The next time you're in the shower trying to recover… recover from an illness, a hangover, or a binge, trying to wash off the ick (that's what I call it) consider this: be ever ready for a *wet moment*. I am not talking about wet dreams or orgasms, shame on you for thinking that! I am talking about a momentary grasp of reason, an epiphany the likes of which cannot be had had it not been for 1. the shower, and 2. the ickies. There absolutely must be both in the equation in order to reach the wet moment.

S + I = WM (Shower + Ick = wet moment)… This is INCORRECT.

S + I do not equal wet moment. S + I = Icky shower. Lol!

There is another variable to this equation, but first let's define the ick.

ick

 interjection

\ ˈik \

Definition of *ick*
—used to express disgust at something unpleasant or offensive icky
 adjective
\ ˈi-kē \ ickier; ickiest

https://www.merriam-webster.com/dictionary/ick

If you ask me ick is something more than an interjection to express disgust. To me the icky thing itself *is the ICK*. To me it's a noun as well as an interjection. Icky things are totally gross, but to me it is more than offensive to my senses. It is offensive, period. Not just to me, but to anything to which it's attached. Ten years ago, when I first began writing this chapter, according to Microsoft Encarta's dictionary there was no ick, but there was icky.

Icky
ick·y [íkee]
adj (*informal*)
sticky: disgustingly and messily sticky
nasty: generally nasty or unpleasant

[Early 20th century. Origin uncertain.]
-ick·i·ness, *n*

- Microsoft® Encarta® Reference Library 2003. © 1993-2002 Microsoft Corporation. All rights reserved.

I appreciate Microsoft Encarta's first definition of Icky... disgustingly sticky. That is the way I had always thought of it. Ick is something that sticks to you, something that has attached itself to you and removing it is not an easy task. Stepping on gum or touching Super Glue is a good example. The worst is Easter egg ink (adhesive) that comes in the itty-bitty pouches inside the kit for making marble swirl eggs. It seems no amount of acetone, alcohol, paint thinner or turpentine can remove the ick and it sticks to you for hours. They really should have a huge warning label on the box. BEWARE OF ICK! Soap as a solvent is our first go-to solution when we have something stuck to us. To get the ick off. It is only natural for us to lather up and spread the suds all over and exfoliate with a loofah! Thank God for soap!

If it were not for the ick we would not come to know our human vulnerability. There would be no reason to clean our houses, cleanse our faces, or take antibiotics for that matter. Ick is more than sticky stuff. Ick comes in all shapes, colors, and sizes, and if you can't see it it doesn't mean it's not there.

Most would agree the absolute worst kind of ick is the kind that pollutes us and can ultimately kill us. Parasites feeding off our flesh, degenerative diseases eating away our systemic functions, alcohol/drug poisoning our liver and other organs, tar-filled lungs, bad cholesterol clogging our bloodstream, to name a few, are the types of ick that can be deadly. Truth is all ick is deadly if we give it enough time to corrode. What about oxygen... funny how it is vital for life yet causes oxidation and corrosion to many other things.

Inasmuch as the ick aforementioned in my equation about the shower and the wet moment is described in a lighthearted manner, I do not mean to minimize its mal significance in and to our lives. The wet moment is achieved in the shower not by successfully washing molasses out of our hair or printers' ink off our hands. The wet moment is profound to washing the toxins out of our bodies and our existence… to becoming ick-proof. Surely soap and water cannot achieve such a feat, but it *is in the bathtub* I've come to recognize its super-sticky ickiness. But more importantly, it is there with the water pouring over my head that I realized the source of it, the implications of it, and the key to unlocking the bond of it.

Many recovering alcoholics and drug addicts join programs in their communities and/or churches to keep the ick off them. To be free from relapse they encourage each other and hold each other accountable. As a group, they team together for strength to be able to accomplish what they could not alone. Victims of illnesses and diseases take medical precautions to reduce or eliminate the ick from which they are suffering. Whether they take medications, participate in physical therapies, or join support groups they are all trying to be free of the symptoms caused by some type of ick.

Ick is not exclusive to physical sicknesses. Mental illnesses and depression are caused by ick. Hatred and murder are caused by ick. Ick is the substance from which our mind and body dissolves. Ick is excrement, trash, pollution, bad bacteria, toxic chemicals, ferocious fungi, and hungry microbes, such as viral parasites. All waste and everything which causes decomposition. Poison. What's more, ick is the substance from which our homes fall apart, our earth shakes, our solar systems diffuse, our universes collapse. Ick is the crud that eats away our flesh, our sanity, our heart… not just ours, but all matter including our planet. Icky stuff builds up, like dust and dirt or grime. If we don't clean our homes the ick will start to cause the walls and floors of our houses to decay. They say cleanliness is next to Godliness. This is not an overstatement.

In our bodies the ick builds up until it becomes dangerously accumulative, and then it can turn into a malignant tumor which will then cause the walls and floors of our body parts to rot, thus explaining our immune system going into overdrive. When I was a young teen I asked a doctor what cancer is. The doctor told me, by definition, cancer is our body rapidly producing cells. I asked the doctor how is it possible... it doesn't make sense? If one who has cancer is "rapidly producing cells" why do they look emaciated? Cancer is rapid cell production? It is an oxymoron of a statement. If that's true then why isn't the cancer patient getting bigger instead of smaller. Back then, I investigated it some more because I was fascinated by this conundrum. I couldn't find a satisfying answer. I noticed today's definition of cancer has been modified to accommodate this question. According to Google, cancer is a disease in which abnormal cells divide uncontrollably and destroy body tissue. This makes more sense. What did they discover under the microscope that they didn't discover 35 years ago, I wonder? Is it possible that cancer patients waste away and become emaciated because the rot from the ick is growing at a greater rate than the patches our bodies are desperately making... to cover up the hole of the rot?

To be clear, rot does not "grow." Rot decays. So, I am referring to the rate at which it decays versus the growth rate of cells rapidly producing. I'm not a medical doctor; I am just trying to make sense of it. Hear me out as I philosophize my ideas. All death can be attributed to the four standard forces, and when unified, a single force. If we get hit by a car, the gravity of the car can kill us. If we get stabbed, the gravity of the knife can kill us. Are you following me? In these examples, gravity is not used conventionally like falling to our death. Fast moving vehicles are accelerated by moving parts, but it is the gravity that is being pushed. It wasn't the electromagnetic force that was pushed, nor was it the weak or strong forces. You see? Gravity can also be "pushed" more slowly. For example, long periods of sitting in one position results in bed sores.

Now with this in mind, picture the ick as fungal mold on food left on the kitchen counter. It "eats" the flesh of the food until it's gone. Bed sores are caused by an infection from gravity that will rot the flesh. It eats it. Poisons, such as animal venom and or toxins eat at the molecular level.

Whether they are classified as peptides or proteins or whatever, something is being eaten.

Now consider cancer's rapidly producing cells. Could it be our body's immune system's counteraction of the rot within us? Wouldn't it explain why our body's make tumors to contain it? The rot? The ick. We've all seen pictures of tumor's dissected to reveal all manner of waste inside. Ick. It makes sense that our bodies which are retrofitted with a defense system (immune system) to fight the force of ick taking bites at us would contain it somehow. Like containing a wildfire. Or to quarantine a virus... like a cordon sanitaire of the rotting ick collected in our body causing cancer. Unfortunately, cancer wins the race too much of the time.

Whether the ick is in the trash can, in the basement, in a landfill or in our bodies, its destructive nature is found everywhere and in everything.

Corrosion as a spreading disease,

Erosion as contagious as the flu,

Radiation as dog with fleas,

Irritation, no doubt a wearing on you

Amputation to stop the rot,

Annihilation to start it.

Fusion, the cure, God's ultimate plot

Ahh, and cohesion... the molecular band aid that fits like a mitt!

Titration to make it tolerable,

Filtration to slow it down,

Intrusion to make it vulnerable,

Illusion to disguise its contaminating frown.

The ick is a form to take away form. In other words, its substance is corrosive and like weak nuclear force a chain reaction is formed to deform the matter of which it touches. Unless the contaminant is severed from the unspoiled part can the rest of it be saved. Ick will cause the surface of which it is connected to to rot. We find evidence of this in molding bread, bruising apples, persons with gangrene or cancer. We find it in dilapidated houses falling apart from mold (like a termite infestation). It is found in all things. All matter is cursed with the icks of life.

Ick is moved by weak nuclear force and strong nuclear force. Ick is moved by all four forces; what else can *move* it? Ick is an eater. It eliminates. It eats us. Not just us. So it is that all of life is a process ultimately of elimination. Everything we acquire will return to dust, including us. All we accumulate culminates to a climax… to an apex after which the rest is downhill. We can watch the petals on a rose deteriorate or witness man's hand wrinkle with age. By the same token, on the other side, we witness man's hand in explosions, implosions, or executions. We also witness man's hand in preventative measures by creating levees, dams, and anti-aging wrinkle cream.

The process of elimination, man-made or otherwise can be seen everywhere, in everything. It happens "naturally" at the hands of Mother Earth or Father Time, figuratively speaking, and "unnaturally" at the hands of homicidal maniacs or mad scientists, literally. Whether it be from fungal spores or a bullet its main purpose is to steal, kill and destroy. Our universe is no exception.

The 2nd Law of Thermodynamics is called entropy. It is defined, in part, as a "thermodynamic quantity representing the unavailability of a system's thermal energy for conversion into mechanical work…" see Google.

Entropy is a T-shirt equation, and a concise one at that:

ΔS universe > 0

This means the net change in total entropy in the universe is greater than zero.

Entropy. Lost Energy. It happens, and it happens at a rate greater than zero. To understand this more appreciatively, let us first examine the 1st law of thermodynamics.

ΔE universe = 0

The Greek letter delta Δ is used to say, "the net change in" and E= total energy (in this case of the whole universe)... that leaves us with 0. The equation says the net change in the energy in the universe is always equal to zero. It stays the same. It doesn't increase or decrease. I recall being taught that "energy cannot be created nor destroyed but can only change into something else." Campfires are best examples; the energy in the wood is not being destroyed by fire but rather cellulose (the fibers in the wood) is being transformed into the chemical carbon (the charcoal) and in the process of this change much of the energy was transformed from solid to gas in the smoke (but the smoke itself contains soot, which remains a solid (in small particles)), but you get the idea; solid to gas. Basic chemistry. The campfire serves to explain the discovery made by Rudolph Clausius in 1850 of the 1st Law of Thermodynamics. Clausius gave us this equation to illustrate the natural process of energy conservation. His 2nd law of entropy seems to work against it.

1st Law: ΔE universe = 0   vs 2nd Law: ΔS universe > 0

Whereas E stood for Energy in the equation for the 1st law, now S stands for entropy in the next equation for the 2nd law. The 1st law conserves energy and the 2nd law appears to eliminate it. Yet, although it appears that way, entropy is not the same as energy. Entropy is the increasing loss of energy in thermodynamic systems. Entropy in the universe is greater than zero. Things are "breaking down" in the universe, not building up. Well, uh the ick may be building up, but that's what's causing the breakdown, for goodness sake! Thermodynamics is the science of heat and how it affects things.

Dictionary.com has a simple definition:

thermodynamics

> [thur-moh-dahy-**nam**-iks] *noun* (*used with a singular verb*) the science concerned with the relations between heat and mechanical energy or work, and the conversion of one into the other: modern thermodynamics deals with the properties of systems for the description of which temperature is a necessary coordinate.

We live in a world affected by heat in machines and in our natural world. If you ask me, entropy basically tells us that the *whole material universe* is in a process of elimination at random rates affected by heat conversion depending on internal and external temperatures to the system. It seems we (mankind) have created complexities to increase this rate. Complexities not just found in mechanics. It's obvious ick comes in many forms, some of which are corrosion, erosion, radiation, irritation. It's all a form of pollution. Entropy proves the process of elimination is in everything. It's not just "mechanical loss," well, unless you can envision all of life as a mechanical system, which, by golly, I guess you could!

I recommend the book Entropy by Jeremy Rifkin. I love his watershed metaphor! *In 1987 Rifkin was criticized by Morowitz (from Yale Univ) for not keeping entropy tied to "equilibrium thermodynamics" and accused Rifkin of applying physics to social sciences. Morowitz is upset about Rifkin not keeping this 2nd law within a closed system of a machine, but Morowitz is referring to a different law altogether (see Zeroth below). Furthermore, isn't the universe itself a closed system unto itself and everything within it cosmic and natural machines? I say, in order to come to a grand unified theory, we must apply physics to EVERYTHING! We must think outside the box, right? We must think outside the *circle* of routine conventions.

Rudolph gave us the first two laws, but there are two more. Zeroth Law of Thermodynamics given to us by Ralph Fowler tells us that if two thermodynamic systems are equal to a third system then they are in thermal equilibrium to each other. Cool name, Zeroth! The third law of thermodynamics has to do with absolute zero. At absolute zero kinetic energy is totally absent, and temperature is totally absent, and so is entropy. As fascinating as these laws are, they're for a discussion at another time.

We will concentrate on the 2nd law, entropy. ΔS universe > 0

While entropy is most commonly thought of as heat loss, like a hot cup'o tea losing its tongue-scalding bite as it cools off on the coffee table and converting some of its heat to the saucer, I think of entropy as the one doing the biting. Entropy is a tricky concept, "they" say, and if you try to figure it out from Wikipedia, don't bother, they make it more confusing than it needs to be. *For the record, Wikipedia is a great place to start any research, but it is not the best place end it. I know I quote them a lot, but believe me, I do it on purpose, not because it is my favorite point of reference, but for its ease of access to general explanations of most things.

The campfire was an ideal example for the first law, but for the 2nd law we need a more closed system, because God forbid we use an "open campfire" to explain entropy. If we did it would sound ludicrous. On second thought, sounds fun. Let's try! The 1st law tells us the energy in the wood of the campfire was not destroyed, but the 2nd law would tell us some of the energy was lost to "the universe." In other words, during the chemical conversion of cellulose to carbon the universe took back some of the energy. It stole it. It would have been the fire's fault! The fire that makes heat caused it to lose heat. Whaaat!

Science does not teach it that way, although it would be "cool" to pick that example apart. When I first learned of entropy it was strictly energy loss in a "closed system" of mechanical work caused by heat. The way I was taught back in the day would explain the campfire this way: the amount of energy in the wood (the mass) itself, if converted to fuel would have had unrecoverable spills (like a glass of milk), and the unrecoverable spills would be measured as entropy. I would have been taught that the spillage was because of the heat loss caused by the conversion. But, I say the spillage is caused by the shaky hands of the universe. The shake of thermal vibrations. The vibrations of radioactivity. The teeth of the slow-moving nuclear force. I know I am making a satire of all this, but I just can't help myself because it is so much fun to play with words!

Seriously, science would have used a computer or a machine as an example to explain this law and their definition is hard to understand. To continue with our campfire illustration an indoor fireplace will do. The purpose of the fire would be to heat the home, right? The home will be our closed system, ok. Entropy says some of the energy would be lost due to leaky windowpanes! It's really that simple. Entropy doesn't have to be a cumbersome concept. The fuel input (energy) is measurably less than the desired energy output. *Kinda like windmills as a source of alternative energy! I don't know why we bother. It costs more to maintain them than the energy they generate. So much for conservation!*

Entropy has four definitions according to dictionary.com:

1.*Thermodynamics*

a. (on a macroscopic scale) a function of thermodynamic variables, as temperature, pressure, or composition, that is a measure of the energy that is not available for work during a thermodynamic process. A closed system evolves toward a state of maximum entropy.

b. (in statistical mechanics) a measure of the randomness of the microscopic constituents of a thermodynamic system. *Symbol*: S

2. (in data transmission and information theory) a measure of the loss of information in a transmitted signal or message.

3. (in cosmology) a hypothetical tendency for the universe to attain a state of maximum homogeneity in which all matter is at a uniform temperature **(heat death)**.

4. a doctrine of inevitable social decline and degeneration.

We just covered the thermodynamics explanation of entropy with the campfire lesson… as I understood it from back in my day. Jeremy Rifkin's book expounds on entropy as defined in number 4 (above) relative to social sciences and we will skip over points 1 and 2. That leaves us with number 3, the cosmological definition. This is the description from which I will delineate my hypothesis taking into account the other definitions as contributory.

A machine in a factory is at work doing its job, but the heat caused by friction of its moving parts wastes energy it could have used to do its job. In this case of thermodynamics, entropy was the amount of energy lost due to the heat of the mechanical system. Entropy talks about the energy. What about the machine? Its wear and tear? Does anyone care about the machine itself? I'm not talking about the "care" of machines. I know we install fans and cooling systems to prevent over-heating. The friction and heat will cause the machine to eventually wear out. Plenty of studies have been conducted to test the strength of merchandise before it hits the market. If you ask me, the same measure Clausius used to determine the thermodynamic quantity of energy loss in mechanical systems can be applied to the deterioration rate of the system too.

Considering everything, I hypothesize that the loss of stamina in natural and mechanical systems is caused by entropic conditions external to the system and not exclusively by the condition of the system generating heat. I am referring to systems as it decays naturally under "natural' conditions to be considered and measured with the same rule of entropy. Is there anyone out there who can help me test this theory or point me to a test already conducted?

Like I said, ick is in the air. There is no way around it. The amount of energy may remain the same as the 1st law tells us, but all matter is in the line of fire of ick's eliminating power and given enough systemic cycles and/or exposure to time entropy will win. There is enough ambient heat in the air to cause breakdown. That is why ancient scrolls and papyrus are stored in ideal thermal conditions. But has anyone measured? Has anyone counted? Is it a countdown to zero?

I believe this loss of energy is caused by super slow nuclear moving ick in the air. Heat eats! This axiom does not need to be tested because time itself is the test. Why is it that we wear sunscreen! The radiation of the sun is not the only ick in the air. Natural and manmade pollution and other man-caused particulates infect the air. Would my hypothetical "entropy in the air" still be measurable if the air were perfectly pure? If mechanical systems were impermeable? As it is, entropy tells us that our universe is stealing. It's greater than zero. Ick is in the air and it has teeth!

The 2nd law disproves order out of chaos. Or does it? Professor Dave explains how order out of chaos is possible. He says, "entropically unfavorable processes can be spontaneous at low temperatures if they are energetically favorable." And guess what can do this? You won't even believe it. Check this out: https://youtu.be/8N1BxHgsoOw

Soap! Soap suds! That's right. I can't make this s**t up, lol!

The loss of energy should not be limited to energy loss from the heat of the machine's own mechanical system when considering the law of entropy. Heat in the air contributing to the eventual mechanical breakdown of the machine should be counted in thermodynamics. Whether the heat is caused by natural or artificial light, radio waves, chemicals, and all pollution, it is contributing to the thermodynamics of our homes, communities, and planet. Entropy is an invisible eater inside and outside of the machine whether the machine is turned on or not. May I recommend a 5th law of thermodynamics? One to include the ick in the air! Entropy should not be exclusive to closed systems as physics traditionally teaches, and I don't think Rudolph Clausius meant for it to be exclusionary because his equation included the entire universe... ΔS universe > 0

The 1st law says that our total energy will move on long after we're dead. Essentially, the 1st Law of Thermodynamics gives up hope for the next generation, whereas the 2nd law takes it away. The 2nd law says the ICK is coming to getcha!

The black hole. We all know what it is... it's a swirling cosmic vacuum, from no one can escape its whirling clutches if you move too close, right? Beyond the Event Horizon, right? Its gravity is too strong to countervail. No one or no thing can withdraw from its pull. Not even light, we're taught. What else in life resembles a black hole? A tornado or hurricane? A whirlpool? A flushing toilet? An auger or a simple screw take on this corkscrew shape. Even things enormous like a galaxy! On the flip side, consider how small we can go... molecular... as double helix strands of DNA coil around each other, and even smaller as electrons orbit around a nucleus of an atom. It appears that so much of life as we know it is helical. Many things resemble the vortex of a black hole. But do all things behave the same as a black hole? Let's hope not!

Chapter N1NΣ

Nine

Some of you may recognize what you see in diagram 37 (next page). The basis of sacred geometry. This spiral is shaped by a formula given to us by Leonardo Bigollo Pisano from Pisa, Italy in the 13th century AD. He is well-known today as Fibonacci, short for filius Bonacci which translates as *son of Bonacci.* See Wiki.

As I mentioned in chapter six, zero's place in the western world was rejected until Fibonacci introduced it. He published Liber Abaci which means book of abacus/calculations, and in his book he used Hindu-Arabic numbers because Roman numerals were insufficient in mathematics. It wasn't until this son of Bonacci dropped the clunky letters of Roman numerals when others began to take notice.

Thank you Filius Bonacci for bringing us zero! And an encore for bringing us the Fibonacci Sequence! This formula is a sequence of numbers strung along to infinitude. They are attached to one another by the sum of the preceding number.

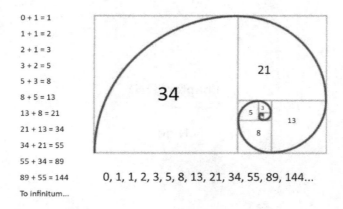

0 + 1 = 1
1 + 1 = 2
2 + 1 = 3
3 + 2 = 5
5 + 3 = 8
8 + 5 = 13
13 + 8 = 21
21 + 13 = 34
34 + 21 = 55
55 + 34 = 89
89 + 55 = 144
To infinitum...

0, 1, 1, 2, 3, 5, 8, 13, 21, 34, 55, 89, 144...

37. Golden Ratio

If you look to the left of the illustration (figure 37), you will see the code of the sequence beginning with zero. Do you see the pattern? 0+1=1, 1+1=2, 2+1=3, 3+2=5. Simple enough, right? How do you get the spiral, you may wonder... from here we have to go back more than a thousand years to Phidias, the most famous ancient Greek sculptor who applied PHI to the construction of the Parthenon in Athens.

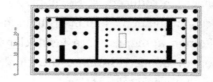

38. Floor plan of the Parthenon

Phi (pronounced fee) is the 21st letter of the Greek alphabet, Φ, and is known most popularly today as a name, in part, for fraternities and sororities (ie, Alpha Phi, and Chi Phi, etc.). Phi is also used as the symbol for the Golden Ratio which Phidias established as the most pleasant dimension when building rectangular structures. The ratio for length to width had to be 1.618. Although the "Golden Ratio" was not coined until the 1800's, and phi was not coined until the 1900's, the dimensional proportion was well esteemed by the Greeks in ancient times. Although we cannot be 100% certain that it was Phidias who discovered this ratio, Historians point to him as the father of the Golden Ratio. This ratio has been around for a very long time, but it is known today by modern pop culture as sacred geometry.

Through the ages, this ratio of 1.618 was revered. From the chronicles of Leonardo Davinci's many drawings from the 15th into 16th centuries include Luca Pacioli's "De Devina Proportione" translated as the Divine Proportion. See https://www.maa.org/book/export/html/116816

For perfect balance and beauty this ratio was applied to DaVinci's famous painting, The Last Supper. This ratio was not confined to the rectangle alone. The ratio goes outside the box. Hehe. If you follow the ratio, you can create all kinds of shapes and move in more than three dimensions.

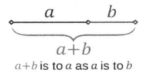

$a+b$ is to a as a is to b

39. Golden Ratio

Just like the Golden Ratio, the sequence of Fibonacci followed the same divine proportion, the whole (a+b) was always divided by the same ratio or increased thereof by the same ratio. The spiral has become the most loved pattern the ratio creates because of its beauty... because of its natural helical appearance in, say, flowers and shells. The spiral is mesmerizing. The ratio can certainly be used in any shape you choose, a rectangle, as it first was, or even on a one-dimensional line. But the ratio takes on more beauty, yes, as a spiral, but more importantly, it becomes more intrinsic to function as we observe this ratio in three + dimensions. Now we're moving.

People are catching on to the natural order of helical motion as the premier function for things which apply to us in everyday life. Ploughs designed to move dirt in a vortex has proved to be more efficient with less resistance than traditional ploughs. Like a mole, the plough moves the soil centripetally. Another example of modern-day inventions to copy nature is the coiled water pipe.

Water moving spirally keeps water from going stagnant. It moves the flow of the water in a helical motion thereby preventing the growth of algae and mold, *the ick*, which would, otherwise, contaminate the water. See the works of early 20th century Austrian-born forester and inventor who "comprehended and copied nature:" https://www.youtube.com/watch?v=yXPrLGUGZsw

This link to the well-done documentary showcases the ingenious works of Viktor Schauberger, from his unique logging flumes, centripetal plough, helical pipes, and his vortex machine, the Repulsine, as well as other vortex appliances.

Viktor Schauberger built the Repulsine, which became legendary after WWII, when it was touted as the first flying saucer, according to the documentary referenced above, when the machine levitated off the work bench. The Repulsine was built to generate electricity, but consequently the vortex movement within it, as rumor has it, caused it to fly! Schauberger's machine had ring rotators and a central axis. His vortex machine was a torus model variety, intended to naturally rotate in perpetuity, so it is not too hard to imagine it repulsed gravity as it spun. Fans of the Repulsine hoped it could supplement the motor of airplanes and submarines as a propulsion device, thereby reducing the need of petrol. Even with his patents, unfortunately, it never lifted off. The last Repulsine prototype sat unused in Austin, TX at the Institute for Advanced Studies where they also run experiments for NASA in research of cold fusion and zero-point energy, until it was returned to Austria where it is now.

Viktor's son, Walter, decades later, built a hyperbolic funnel, from what he discovered after playing the Monochord. Like a harpsichord (which my grandma played along with the piano), a monochord is a much simpler instrument devised by Pythagoras according to the documentary noted above. Who knew! Pythagoras used the monochord to calculate math ratios which may have led him to his famous theorem. Wow! The single-string instrument has a bridge which can be moved along the chord to change the frequency of the tune produced by the one string. In ancient Greece, this was used to study the marriage between music and math. See Wiki for Monochord.

Walter found a pattern in sound when he moved the bridge by one half. When you cut (block) a string in two the frequency doubles and it increases the octave twice as much. When you cut the string by a third, the frequency is three times as high, and quarter is four times. He then applied these numbers to the Cartesian Coordinates. From the following illustrations (figures 40, 41, and 42) you can see how Walter came up with the dimension of his cone. First, on the graph, he discovered that this pattern drew a two-dimensional (2D) hyperbola (see figure 40), and when he rotated the hyperbola into a third dimension, he got his hyperbolic cone (see figures 41 and 42)! He called it the Tone Tower derived from what he called the Law of Sound.

It astounds me that the fundamental geometrics the Cartesian Coordinates provide to us can result in such discoveries. And more astonishing, Walter's Tone Tower, derived from the coordinates established by music (acoustic energy) from the simplest string instrument unbelievably brought Walter full circle to his father's invention of the helical water pipes. Extraordinary! His Tone Tower became a hyperbolic funnel the likes of which his father, Viktor, hoped to achieve with his coiled pipe creation. He would have been so proud had he been alive to witness it. The inexplicable coincidence of these two inventions by father and son blows my mind.

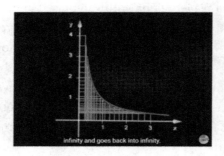

40. 2D hyperbola from doubling tone frequencies

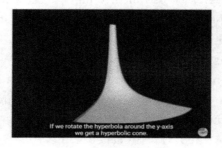

41. Rotating the 2D hyperbola to make a 3D model

42. 3D Hyperbola nearly whole to form the Tone Tower

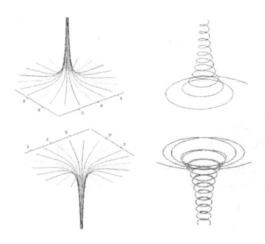

43. Tone Tower to Hyperbolic Funnel which produces natural double helix spiral motion of water

When Walter converted his Tone Tower (by turning it upside down) into a hyperbolic cone and found that using it as a funnel for water was as effective, if not more, than his father's corkscrew pipes, I'll bet it blew his mind as well. Like his dad's pipes, the water moved in a spiral, but unlike his dad's pipes, this funnel of water moved in a mathematical double helix, naturally (see figure 43). It was fine tuned. Literally. He also discovered that the water moved in longitudinal vortices and found an oval shape when the cone was cross-sectioned, showing the path of the water as it moved its way down and around the funnel in its spiral and elliptical descent. Walter likened the cross-sectioned oval/egg shape to Johannes Kepler's ellipses formula from the 17th century. Walter's funnel (and his Law of Sound) he dubbed, was a link between the harmony of Pythagoras and the astronomy of Kepler's Laws of Planetary Motion, so he merged the two into the Pythagoras-Kepler-System, aka PKS, which in the aftermath stood for Pythagoras-Kepler School which he founded in Engleithen, Austria. Walter's son, and Viktor's grandson, Jörg Schauberger, in this family's third generation, is carrying the baton of this research by resurrecting the works of his grandfather and teaching at the school his father created to further bring to life inventions copied from nature.

Kepler's formula of the elliptical planets orbiting the sun also mentioned the Divine Proportion and was, no doubt, considered in his research regarding heliocentricity. In 1543 AD it was Nicolaus Copernicus in his publication De Revolutionibus Orbium Coelestium (Latin for "on the revolution of heavenly spheres" see Google), who caused western scientists to finally take seriously the idea that all planets revolved around the sun. Prior to that it was widely regarded and accepted that the planets revolved around the earth in a geocentric model given by Ptolemy. Johannes Kepler rang the bell, so to speak, on Copernicus' assertions, thereby changing the then worldview of our solar system. It was made official. Thereafter, it was a science not to be questioned. We (earth) are not at the center of the universe and everything does NOT revolve around us, *so we're told.*

This new "Divine" order of things ushered in a new era for science and math, I'm sure, and began the journey to enlightenment for many western scientific thinkers. The Golden Ratio would not have been realized had it not been for the Hindu-Arabic base 10 decimal system, and have neither would algebra, calculus, geometry nor trig. Nor math as we know it today. But we are entering into a new era of arithmetic. Or, as I like to call it... *arithmystic.*

A new age of thinking and enlightenment for many who are dressed in cosmic fashion as they celebrate the dawn of Aquarius believe the next 2150 years will be ushered in by a renaissance of quantum-unified reality. This is my way of saying that modern pop-culture is pushing a paradigm of unification. You've heard it! One world government. One religion. One mind. There are many schools of thought, but they all have one thing in common. Unification. A cosmic consciousness through collective thoughts allied in positive vibrations is the new age religion. Being one with the universe. All to what end? To achieve utopia or come as close as possible, what else? Live in harmony as a unified species. Sound good? *Of course, who would argue?* Is this a righteous idea if it means we will eventually unify with other species, as well? Without God? How will we achieve this new order of things? Out of chaos? Surely, we seem to be approaching mass chaos. In the streets. In schools. In congress. In families and even in churches. Chaos is expanding globally. World War III looms as does a world-wide pandemic. Novel Coronavirus may just be the beginning. A global catastrophe is presented in every imagination on Hollywood screens!

Fyi, the vernal equinox does not determine ages, only seasons. Ages are calculated by earth's gyroscopic precession of approximately a 25,800 year cycle divided by 12 zodiac constellations (see en.wiki, Age of Aquarius). The gyroscopic precession is the measurement of earth's wobble on its axial rotation using the circum-polar stars as a gauge. The sun is not set to leave Pisces for another 1080 years or so. We are hardly at the dawn of the Aquarian age.

Entropy.

I'm trying to weave this all together. Entropy. Ick. And the black hole. Hang in there.

Arithmystics will tell us they can prove with math that the solution to all the above is in unification. In a single (hive) mind. A singularity. A growing idea of this can be found in vortex mathematics.

Vortex math was used and experimented on early in the 20th century. Nikola Tesla is probably the most recognized scientist of the era who was fascinated by it. His famous proclamation was, "If you knew the magnificence of the three, six and nine, you would have a key to the universe." Today, vortex math has gained momentum due to the Torus model of vortex energy as the shape of the universe.

Shape of the universe? A black hole! God forbid!

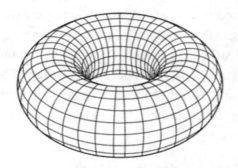

44. Torus vortex spiral. See a torus in motion at
https://tenor.com/view/torus-vortex-spiral-gif-14087761

The torus model was first proposed by Alexi Starobinsky and Yakov Borisovich Zel'dovich in Moscow's Landau Institute in the 1980's. About a decade later, in the early 90's, Marko Rodin enters the vortex scene. Since then he has claimed to have discovered the secrets of the Torus (I mentioned Marko in chapter two). He found a number pattern and heroically insisted it was the discovery of the Grand Unified Field. This number arrangement is as fascinating, if not more than, as the Fibonacci sequence. This number pattern reveals the toroidal shape we see in figure 44, and his claims echo this doughnut-shaped black hole is the template of our universe and much more. His claims also echo the famous quote from Tesla regarding 3, 6, 9. However, if you research on-line you will find serious mathematicians are not entertaining Marko's self-proclaimed idea that his discovery is mathematical. There is no formula to a solution, after all; it is merely a number pattern. This pattern, as magical as it appears, does not prove anything, say many opponents. Here are two must read articles opposing vortex math:

http://www.goodmath.org/blog/2018/01/02/zombie-math-in-the- and vortex/

https://www.theproblemsite.com/vortex/

Still, Marko Rodin's ideas are spinning more and more followers into his fabric of reality as he trumpets his discoveries, one of which being the source of non-decaying spin of electrons. This claim remains to be seen, but the pattern of his grouped numbers is undeniably remarkable. To make his Rodin Coil you must follow this sequence: In figure 45 you will see the numbers 1-9 placed in a circle.

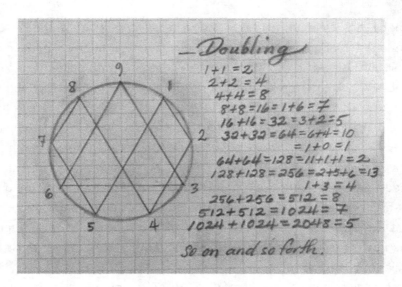

45. Vortex Math

The placement of the numbers is deliberate and create the "flow of energy." The patterns reveal each number to be alive, according to Rodin, and are significant to everything. *Significant to the universe, to free energy, to curing cancer, to each other and to itself.* Most of his claims remain to be proven, but I personally know one thing's for sure; God's got the whole thing in *His hands*. God is the ultimate scientist. The ultimate mathematician and the ultimate authority over all things including numbers. Rodin says each number itself is more than just a quantity. The "quality" of each number, he boasts, "has finally been revealed," and within these number patterns the secret of creation has been uncovered. The numbers are connected and divided. I.e., 1, 2, 4, 8, 7, 5 are connected and reveal an infinity symbol, geometrically abstract with its six points. Three, six and nine, are divided into a 2nd group.

I recommend watching this video to fully understand and appreciate the sequence: https://youtu.be/OXbVZc10lnk by know-how

Here are more references: https://youtu.be/MG_2IRAHeh0 by Knowledge is Power - Gary Lite

Indeed, the connection of the numbers 1, 2, 4, 8, 7, 5 is mind-blowing. In addition, the polar dynamics between three and six are awesome. And the yin yang symbol with nine controlling the polarity as a holy trinity. Wow! These sequences are truly out of this world. I, too, believe, this order of numbers to be significant in their places in existence, but unlike these followers of Rodin, I will not lean toward a cultic assembly of one of many eastern religions. Moreover, this discovery does not have me 2nd guess the formation of the universe and cause me to rest my soul on a foundation of beliefs contrived by one man less than 150 years ago...

Rodin claims (via his follower, Gary Lite) after studying all the "worlds' great religions" he settled on the Baha'i faith founded by Baha'u'llah in 1863 because the number 9 is deemed the number of God, ABHA, decoded from Baha'u'llah using Arabic numerals. I had never heard of the Baha'i faith, but it seems to have a link to Zoroastrianism. Since Rodin's conversion he has taken his "secret sequence" discovery and turned it into a "secret cult" if you ask me.

Rodin's claims of having found the Grand Unified Field of the universe is attractive to his growing number of followers and they are eating up the idea that the number nine is divine.

Nine is also considered sacred by other schools of thought. Google it. There is no shortage of information about nine and why it's so "sacred." Nine is a Motzkin number, nine is an exponential factorial. Nine is this and nine is that. It's true, nine is special.

*Fun Fact: I personally, think it's pretty neato that **multiplying any number by nine, will give us a sum from which if we add together all digits, we will always end up with the number nine. I.e., 3x9=27, 2+7=9, or 9x9=81, and 8+1=9. In the case of 9x75=675, we would take an extra**

step to add until we reach a single digit... 6+7+5=81, then add again 8+1=9. We would follow this same formula for multiple digit sums until we reach a single digit, which will always be 9. For more rules of nine, see https://www.washingtonpost.com/archive/1998/05/13/the-many-mysteries-of-the-number-9/dc65f743-0f71-4b16-8420-8519bc7fa445/

Pretty cool. But not cool enough to risk my soul or eternal destination. Rodin's right about one thing: "the quality of each number has finally been revealed." Indeed. Let's examine nine from Rodin's model. It sits at the top of the circle and is the control of three and six. It lines up with the center of "infinity" and, therefore, must stand alone as unique. *It must be heralded as the God of all numbers. In the numeral kingdom, Nine is king!*

But wait, what about the ace? Where is the ace? Aces trump kings, don't they?

The ace is the *circle, itself*. It carries the whole system. Without it the entire model collapses. Is it any coincidence that this model conspicuously bears the missing zero as the cupbearer of the entire sequence? The Zero, here represented as the circular model of all its parts, is rightfully in its proper place, encircling the whole set. See image 45.

As we saw in chapter six, zero is a gate. All numbers 1-9 can be multiplied infinitely and easily from a zero placeholder, allowing it (any number) to enter a new quantity. In chapter six we also saw from the system of coordinates that zero is aptly placed in the center. It tethers all functions unto itself. Like the Cartesian Coordinate structure, the Rodin Coil will collapse without the zero. In geometry, the zero is the central point, in vortex math the zero is a womb carrying the numbers, respectfully.

As you can see, it is a no brainer. It doesn't take an entire dissertation to make this argument. The pictures and numbers speak for themselves.

Nine is just a number. I agree, nine has unique properties, and stands alone for many reasons, but it is, nonetheless, contained within the circumscription of zero. There is no way *around it*, hehe.

Nine, I believe, may represent something more sinister than we can imagine. Just like the shape of it, the vortex seems to be revealed in its function. The spiral, as beautiful, and with "semblant" life-making properties as it contains, has a dark side. Allow me, please, to express my thoughts in conjecture to this notion. Like in our number system with positive and negative integers, life as we know it contains matter and antimatter. Light and dark, day and night, up and down. Good and evil. For everything there is an opposite. Happy-sad, good-bad. Sucky-rad!

By the end of chapter six we saw zero established as a place holder and a gate, but I asserted the possibility that it also serves as an anchor and vortex, as well. If we look at figure 46 showing Quadratic Functions of the Vertex Formula, one is for positive (+a) and the other is for negative (-a). The parabola opens downward for negative and upward for positive. Either way, the image indicates the functions are explicable with points and axes to measure its value.

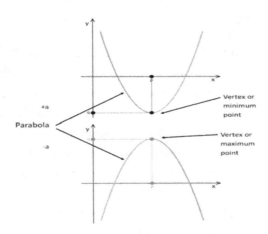

46. Quadratic Functions. Positive and negative parabolas

Positive and negative vertices are illustrated here as maximum or minimum points to define the function of the curve as either forward or backward. Plus or minus. In the same fashion you will have a positive and negative function for every dimension in between. It is easy to imagine this same principle for everything. Positive and negative functions are found in everything. It is congruent to quantum particles having an anti-particle. This is not a designation exclusive to math only; all particles are life forms. All abstract concepts are descendants of life. Literally, every-*thing* has an anti-*thing*. We have light and anti-light (or darkness), fast-slow, happy-sad, salt-pepper... you get the idea. You can have a positive spiral and a negative spiral. One points up and the other points down. Zero, itself, is not a vortex and does not serve as a vortex. Good, bad, or otherwise. Zero, unlike numbers, does not have a negative 0 (-0). But, like a gate it allows the functions to enter into positive or negative dimensions. As we saw in chapter six, zero serves as a gate to inter-dimensions too. I would postulate that zero acts as the *funnel* in Schauberger's cone, and the water represents the digits (numbers 1-9). I would agree with the functions of all the numbers in Rodin's Coil, save for the number nine. And I would add that 1, 2, 4, 8, 7, 5 represent matter (positive matter) in our terrestrial existence, where six and three function as poles but I would disagree that they embody space and time the way Rodin sees it.

I see nine as an auger cutting through space, letting "loose time" leak in. Nine needs 3 and 6 to function. To rotate, so to speak. Nine alone is not holy. Nine joined with 3 and 6 does not make it holy. Nine is the Dr. Jekyll/Mr. Hyde of numbers. Can't fully trust it. I see nine as the radioactive substance which must be dealt with care and delicate hands. Like a snake charmer approaching the poisonous coiling cobra. Like a CDC Lab Tech approaching the Ebola virus. Like the oncologist approaching a malignant tumor with his scalpel. Like the engineer or physicist approaching uranium ore. Like the negotiator approaching a psychotic maniac hellbent on blowing up himself and everyone around him.

Nine is the rogue number. It pulls all the others into its grasp. Like a black hole. Once you're sucked in there is no escape. Nine is the ICK. It is leashed to zero, but like a mad dog it wants to break its boundary and pull out of its stable limitations. If given the opportunity it will; leaving a wake of destruction in its path. A trail of blood. Like an auger drills into ground, nine drills into flesh. It's a molecular screw punching holes into mitochondria creating black holes of cancer which our bodies try to patch up like a band aid by damming it like a beaver. Before we know it we have tumors the size of beaver dams because instead of closing the hole we just keep feeding it.

Nine. How can it be so terrible if it is one of the numbers necessary for life? If God made it? Our numeral system would not be complete without it. To that I'd say, nine is important and serves an important function as do all numbers, but this one specifically can wreak havoc if not managed.

Is it a strange coincidence that the number nine is revered by secret societies known today to be Luciferian in nature? Freemasons consider nine to be the number of human immortality. Cabal says it is linked to the divine will. Nine has been touted as the premier number in "sacred geometry" by *arithmystics* dating back to 3000 BC in India. Is it a coincidence that CERN's Large Hadron Collider (LHC) has superconducting magnets all along its ring which measures 27 (9 x 3) kilometers? Is it a coincidence that both Stephen Hawking and the pope have been quoted saying that the LHC could open the gates of hell? Is it a coincidence nine is linked to the dragon in Chinese culture? The dragon being the serpent! The serpent being the devil. Is it a coincidence that nine in its phonic state means NO in other languages and dialects? Nein, na, ne, no, nem, nei, etc. Ha, nine is negative in its positive state. It sounds bad and it looks bad. Lol! I don't know about you, but I wouldn't want to worship something which depicts itself as a black hole. Or used to put people in a trance, for that matter. I do not see the unity in that. I see the mouth of a meat grinder. A closing in. Looking at it I do not see ascension. It gives an impression of descension. Reminds me of Nine Inch Nails (NIN), The Downward Spiral.

47. Fibonacci spiral, 48. Number nine spiral, 49. Spiral vortex

To see a list of symbolisms for nine check out:
https://www.ridingthebeast.com/numbers/nu9.php

Some may argue and say the spiral looks like expansion. An opening up if you stare from the middle. It depends on your own point of view, they say. The famous Psychoanalytic, Etchegoyen, believed nine was the expression of the holy trinity. Go figure. Whether we stare at it from the center or from the perimeter it is still within the zero. It is encircled by it. It has no form or power without it. Expanding or not, it is tethered to zero.

I am a firm believer in sticking with the simplest of explanations. I prefer to go with the most obvious. The truth is not hidden. It is not a secret. It's not rocket science. You do not have to solve a long equation to find it. It is not revealed by vortex numbers. The truth is simpler than that. The first synonym of "simple" offered by Thesaurus.com is "clean" and that is what the truth is... it is clean. Pure and simple.

It's true the spiral has many wonderful properties and is extremely useful when we apply this shape to tools and to engineering improved methods of delivery of water and energy, and much more. I'm sure we will not hear the end of it. It is an amazing discovery. So amazing it is putting people into a trance for God sakes. This is the one number, I think, which has the power to usher in a new era of technology. A new age of awareness, or unfortunately, for unsuspecting worshipers, a new age of deception. Beware of the vortex! I'm serious. Remember, Dante's Inferno has 9 circles of hell!

No one ever says, "you're spiraling *into* control. Why put faith into something that has so many negative connotations associated with it? If we had to pick one number 1-9 that reminded us of sticky icky ick which one would we choose? Allegedly, early Hindu Arabic numerals had angles in its glyph to match its quantity. One had one angle, two and two angles, three had three, and so on. Nine had nine angles. More angles than the rest. More hooks. Claws. Teeth. To "gab" hold. To stick. To latch on and stay latched on, like Velcro. Ew. Sticky icky ick. You already know how I feel about coincidences. *And now you know how I feel about Velcro. Lol!

9 is a snake! Some may argue that 6 is a snake, after all, 666 is the number of the evil beast as written in Revelation's 13th chapter about the antichrist. Not some, but most will argue, I'd bet, that the number 6 is the true "evil" number. You know what, I wouldn't argue because 9 is an upside down 6, and vice versa, and you know how the devil loves that kinda s**t! As above, so below. Get my drift?

For the reasons already stated above as nine being unique (even though it is connected to three and six in vortex math), we will just accept the fact that people are going GAGA over nine. Not just the gurus and Arithmystics. It's all over the place; I first heard about it on a Ted Talk. Many are calling nine divine and the spiral holy. Check this out by Virginia Swain:
https://virginiaswain.com/the-spiral-is-a-symbol-for-my-soul-journey/

Her spiral diagram begins with "spiritual emergency" as you enter the vortex and ends with "a phoenix rising from the ashes" to represent resurrection. Virginia says she feels the holy spirit and receives a message from the holy spirit when she "walks into the spiral." Then practically in the same breath she says she feels the same "unity" she felt in the spiral as she does when she walks into the United Nations Meditation Room. The UN's meditation room has a black cube at the altar! *If you do not know what the black cube represents, then please research its occultic significance. I will not get into it here. Virginia seems like a very nice lady, but I wonder if she knows she is playing with fire? She goes on about the Phoenix and restoring faith in the UN. More evidence of the Cabal? I will talk more about this topic in a future chapter... playing with fire, the occult and its hand in Babylonian tradition as well as politics and main-stream media and Hollywood and in-your-face Luciferianism.

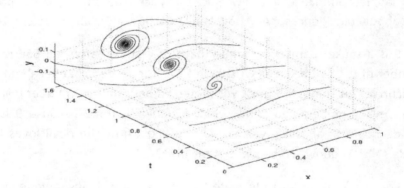

50. Kelvin–Helmholtz instability: the vortex sheet roll-up into a spiral is shown as a validation of the numerical method. The number of point vortices is N 6 1000, ♪ 6 0.01, and ♪ 6 0.1.

Vortices, by nature, are unstable. The spiral has useful functions, and I do not underestimate its ability to funnel water more cleanly then linear pipes, or to dig dirt more effectively than shovels. I do not deny its natural faculty, found in DNA strands, marine shells and flowers, or solar systems, to be intrinsic to life and the beauty it provides as we behold its shape, but the spiral as a vortex machine is dangerous if unleashed. The vortices of a spiral motion we can easily imagine in our mind's eye moves in and out. Back and forth. The negative movement of this phenomenon is as significant as the positive motion of it. It's unstable.

The number nine is special. Super special. So special that we must keep it in check, or it can become a runaway train. When it does we will be hard pressed to stop it. Like radioactivity. Soft flesh doesn't stand a chance. It can drill through it. Punch a hole through space and time? Is it the ionizing energy of weak force? Is it the vibrating mouth with the teeth of millimeter waves? The jaws of parasites and poison? The cause of a rotting hole? The hole cancer cells try to fill? A black hole caused from ick?

What can stop it?

Only something that can *hold* it. Zero, the place *holder*?

How do you get rid of ick? A hot shower might get mud off our body, but how do we get rid of the molecular ick? The mental ick? The spiritual ick?

The key is in zero, "particly" speaking. Now, where did my atomic knife go?

Chapter *Nine* post script: After having written this chapter, I've seen the number nine pop up everywhere and it is usually connected to a vortex, something sinister or suspicious and somehow connected to the occult... fyi. Keep an eye on 9!

Chapter TΣN

Wipeout

I was an animal. Most people would probably listen to their doctors when they instruct you to not put pressure on your leg. Not me. Oh no, I paid no attention to their expert advice. I knew best. I did what I wanted and when I wanted to do it, not considering the repercussions. I recall once having sprained my ankle and it was put in a plaster cast. The doctor instructed me not to put any weight on it for one or two months. As the nurse reiterated what the doctor said she helped me to my feet and handed me a set of crutches which she just sized to my height. The following weekend I was on Mount Crested Butte! Not skiing, mind you. But I was dancing! At the foot of the mountain was a popular spot called Rafters Bar and Grill. It was a ski in ski out place for patrons to grab lunch or sit and drink from their menu of beer on tap. It was also a place for students to gather on the weekends to drink beer and dance to whatever live band made its way to this neck of the woods. This night Pato Banton took the stage. Reggae on the mountain always drew in large crowds. The place was packed and so was the dance floor. If you could go back in time you would find me swiveling to the beat of the snare drum, one hand on one crutch and the other clutched to a beer. By the time I got home that night, my plaster cast was wasted and so was I. I returned to the hospital the following day to get a new cast.

*I said I was on the mountain… not skiing, but dancing. For the record, not because I had the good sense to not ski, but because my foot wouldn't fit in the boot with that dang cast!!!

It was toward the end of another school year and the end of the ski season. I wanted to ski! In another week's time the "Last Day on the Mountain" would be here. The biggest party of the year! During the day and into the late happy hours of the evening! Crested Butte Ski Resort had a tradition which made this day particularly popular, nationwide. On the last day anyone who was willing was allowed to shed off all their clothes and ski buck naked. A parade of snowboarders, alpine and tele skiers, mono skiers, and even kids would parade down the front of the mountain. That year a clip of the parade was featured on the David Letterman Show.

This dang cast was gonna rain on my parade, I would think to myself. I didn't have the willingness or the self-confidence to ski in the nude, and neither did any of my friends. We would be spectators, gladly. We did, however, plan to dress in Hawaiian attire while skiing that day. No one would be wearing their usual pants and jackets. I did want to join in the fun and lucky for me I was able to…

Back in the emergency room the doctor took one look at my shredded cast and shook his head as if he knew how it happened. He didn't even have to ask. After sawing off the cast, the doctor asked me if it hurt because he noticed the swelling was gone. I said no, so he decided to run it through the Xray again. It was a miracle! The doctor was amazed, scratched his head, and released me, cast-free!

Woohoo! I get to ski the "last day!"

On that last day, the sun was out shining brightly... at first. In the morning it was a Blue Bird Day, not a cloud in the sky. By the afternoon, clouds had rolled in and the parade participants were nervous, I'm sure, of freezing their butts off! Lol! A large crowd gathered at the bottom of Red Lady Lift. We made our way to The Artichoke, a restaurant a ski-poles throw away from Rafters. The Artichoke was not a regular place for students to gather. It was too fancy for rowdy students. It was known for their artichoke soup; a puree of artichoke hearts and cream and decadent spices (if I had to guess, lol). I had eaten there once; the soup, a salad, Filet Mignon, and thereafter only went for soup occasionally because, really, it was all I could afford. Today, The Artichoke opened up its outdoor patio for the celebration, equipped with gas grills grilling food at a discounted price, waitresses making rounds with Jell-O shots, and a live band jamming on the upper deck... all with an unobstructed view of the slope from which the "without-a-stitch" procession would make its stark-naked way down the mountain. What a day!

All the glitter! All the glam! I'm not just talking about the decorations or the band. Everyone was dressed to the *nines*! It was like Mardi Gras! Today was a day where anything goes. If you didn't have the balls to go bare, then you wore your birthday suit. Monkey suit, costumes, wigs, props, you name it. It was a sight to behold. What a show! The finale, of course, being the parade of the brave and leafless locals in the buff making turns as they came down the mountain, waving at the onlookers and blowing them kisses. All locals... well mostly... maybe some tourists too.

It was the beginning of the ski season during my sophomore year in college when I took a fall, dislocated my bad knee, and had to be taken down the mountain on a sled. It was kind of humiliating because I wasn't on a difficult slope nor had I just tried a cool trick off some kicker (a jump). It was on a section of the mountain you would never be able to find on a map, because it didn't exist as an "official run." It was not roped off either, so it was perfectly legal for us to be back there, well, not the smoking pot part, but it was okay to be back there skiing. There was a bowl beyond a section of dense trees snowboarders loved! *A topographical bowl*, not a pipe bowl, lol! … Snowboarders loved the bowl but hated the aftermath. Getting out of this area could be tenuous if you didn't keep your speed to gain enough momentum to get you passed a flat bottom back to the lift. Unlike skiers, if snowboarders found themselves without gravity, they would have to unmount and walk it.

That's where I fell. Not in the bowl. Not in the trees. On the easy part where you had to pick up speed if you didn't want to walk. I didn't "catch an edge" which typically only happens to me on icy groomed runs. It happened so fast; my skis must've crossed.

"Dang it! I'm much better than this!" I'd chastise myself. Green, blue, black. Even though this run was not on the ski map, it would have been classified as green or blue.

My friends went to go get help. Ski Patrol EMT's took me straight to the First Aid Shack. They said it appeared I had dislocated my knee. To be honest I do not recall if I popped my knee back in place or if it did it on its own, but I was back on the ski mountain a week later, without having gone to the doctor, right as rain. I got lucky.

Not long after, on one of the best powder days of that year, I wasn't so lucky. It had been snowing for over 24 hours and the fresh powder was light. It was perfect. The snow was still coming down, but the lighting was not flat. You could see just fine. I recall the day being so magical. It's not often you have epic conditions like this! We were making fresh tracks everywhere! My friends and I took our normal route: The Dooby Queen to the T-Bar. It was actually called The Silver Queen (nowadays it's a high-speed quad called The Silver Queen Express Lift) but back then The Silver Queen was by no means an express lift, nor was it silver. It was a three-man lift which never moved fast enough. The reason we would take it was because it would take us directly to the Palma lift, but more importantly, it was a pod enclosure which would shelter us from the wind, so we could smoke our *dooby* easily. The pods resembled black beetles and it always made me think I was boarding a carnival ride at the state fair. The hood was convertible, so if it were a nice day you wouldn't need to pull it down to shelter you from the wind and/or snow. For us, the hood would always come down, rain or shine (or snow-wind-or-shine) at least at the beginning of the lift while we toked our pipes!

From the Dooby Queen we headed straight to the High Lift. I know what you're thinking, but it wasn't called the *high* lift because of *that*. It wasn't so easy to smoke pot on that lift, though I'm sure we did! It was and still is the highest (in elevation) lift on the mountain, so the name is well suited. Back then (and maybe still), the High Lift was just a T-bar that pulled you up to the Headwall two at a time. A slow-moving lift, and you would have to pay attention to maintain balance as your skis or board were/was still in contact with the snow-covered ground, made slick from the tracks of those before you. Or made difficult with deep snow if you had first chair! The Headwall was a workout! No easy way around it. No groomers up there, and it was all straight down, double black diamond, unless you traverse to the west and front of the mountain, you could find a single black diamond. From the front, nowadays, you can hike to the peak of Mount Crested Butte, launch from cornices and ski a variety of narrow shoots and funnels making your way across or over rocky cliff obstacles. Back then, the peak was roped off, and only a few chutes were available on the front. For us, our normal route would keep us on the back side of the mountain. From the headwall we would make our way to the North Face (for which Crested Butte was made famous, I think). To access the fresh powder of the north face you had to take another unconventional lift. This time the Palma Lift. Single file, one-at-a-time. Keep balance as you're pulled uphill, holding onto a single pole with a disc between your legs.

The North Face had a lot of options. Spellbound was a favorite because of its cornices but you had to traverse to get there. I skied with snowboarders much of the time, who hated traversing. Who could blame them... they get stuck having to walk most of the time, unbuckle and carry their board. Angie hated traversing and would avoid it at all cost. She was my ski buddy who was a snowboarder. I've mentioned her in a previous chapter (we went to high school together and shared the same hometown). We would stick to the front of the North Face usually, and the only place to find single black diamonds. I didn't complain. Double black was a little much for me, I'll admit. Steep, scary, and could take the fun out of the run. Unlike most of my friends, who were expert skiers already entering as freshman in college, I had to play catchup. I went from the bunny slope to double-black diamond in one season. Practice makes perfect. I wasn't perfect by any stretch, but I did have a lot of practice. And by this year, my third full season, I could keep up, and sometimes lead the way. And I finally had decent form which increased my confidence!

Speaking of form, by the afternoon, I was itching to go to the front of the mountain to practice on my favorite mogul run. The bumps on Twister. The North Face had bumps too, but they were steeper and more uncomfortable for me. I wanted to be careful because I had recently dislocated my knee. Twister, I loved. The incline was perfect for me. It was a single black diamond run as opposed to double black. I could practice my jumps too! The amount of fresh fallen snow made for a perfect day to exercise moguls and soften landings, I thought to myself. I decided to go the front of the mountain, on my own, because my friends, including Angie, wanted to stay on the Face (the North Face).

I hit the bumps on Twister and stayed there the remainder of the afternoon. The snow was still coming down good making more "freshies" and the lighting was still good too. It wasn't over-crowded, and I found fresh-powder-covered-bumps with each run. What an epic day! I didn't want it to end. I was honing my skills. Moguls weren't my thing, but today, you couldn't tell. I was killing it. The foot (or more) of fresh snow really cushioned the bumps and made it easy on the legs, especially the knees! Snow absorbed the shock of every turn. I seemed to ski better alone. No pressure, I guess. I was happy!

The snow was coming down hard by the end of the day and the lighting was becoming increasingly flat, but it did not discourage me because I was having the best day of my ski life! It was 3:45 pm and everyone was gone. I had the mountain to myself. There wasn't another skier or snowboarder in sight. I told myself if I make a bee line for it I could catch the Dooby Queen before it closed at 4pm, for one last run. I cut left (skiers left) to take International, a run with no moguls, so I could go faster. Faster I went. A friend of mine who had a speedometer clocked us going straight down International at over 50 miles an hour the season beforehand. I was going about that fast! On the right bank of the run was a narrow group of moguls, so I slowed down to inspect. I decided to hit it! I hit a patch of fog too. At the bottom of the bundle of bumps was a cat track. *For those of you who don't know, cat tracks are flat paths made by Snow cat tractors (snow grooming machines) which intersect ski runs. They create jumps onto the incline of the run below it.

As I hit the cat track and was made air-born by it, I saw a poor fellow in my path as I cut through the fog. You couldn't see him until it was too late. How do you turn when you are in the air? *Remember the last time I tried that? During Lacrosse practice in high school? We know how that ended... well, unfortunately, this ends much worse. In this incident, I didn't or couldn't turn in the air, and therefore collided into this unsuspecting snowboarder who appeared to be crossing the slope from the left, moving backward. I smashed into him like a Mack truck and we both went barreling down the mountain, tangled together. Head over heels. Head over heels, again. I mean, head over skis! Until I lost them. Both of them. It was a total wipeout!

"Oh my gosh, oh my gosh, are you okay," I asked earnestly, as we unraveled from each other, eyeballing him up and down to see if he was hurt. If he was bleeding. "I am so sorry. Where did you come from? I thought I had the slope to myself."

"I'm okay, I'm okay," he responded. I was so relieved.

"Are you sure?"

"Yeah, I'm fine," and he snowboarded down and away.

As for me, I was not okay. I probed my surroundings to ascertain the damage. It looked like a yard sale. One ski, over there, the other ski farther down, my hat up there, and my goggles... *oh, okay still on my head, twisted, barely hanging on. Holy cow!*

And my leg! Oh no! Not again! I sat there for a while, waiting for someone. My poor victim snowboarded down the mountain without asking me if *I was okay*. How rude! I thought to myself. I couldn't move my leg without jolting pain shooting from my bad knee causing me to inhale loudly with sputtering beats.

"Ah ah ah ah ah," I moaned in staccato as I favored my leg and moved into a standing position. I didn't see anybody else. I couldn't see anybody else for the fog. I had to get down the mountain myself, because I refused to go down in a sled again. Painstakingly, I managed to gather my belongings and snap my boots back into the bindings of my skis and made my way down the mountain. I continued to favor my leg by not putting any weight on it as I skied down, essentially with one ski. I did my best to avoid turning. Like on the run from my other accident mentioned earlier in this chapter there is a stretch on International that flattens out and, unless, you approach it with enough momentum and speed, you will be trudging... trekking like a cross-country skier. *Ouch, that would require two good legs*. I could not manage such a feat, so I made sure to keep my nose pointed down and take my chances with speed. High risk, but as long as I maintained my balance, no pain. It was worth the risk.

I was so proud of myself. I made it down in one piece and with little to no pressure on my knee. At the base of the mountain I found someone to help me to the First Aid Shack. Two days later I was at my orthopedic surgeon's office in Denver getting X-rays.

"You completely blew your knee." The doctor said. "Your Anterior Cruciate Ligament (ACL) has been completely torn and we will have to take a piece of your Meniscus Tendon to create a new one."

"Can't you just sew it back together?" I asked, not realizing the scope of my damaged knee.

"There's nothing to sew together. You completely shredded it and your body dissolved all traces of it. It was already compromised from your accident four years ago (from Lacrosse)," my doctor reminded me.

*For the record, it was due to that Lacrosse accident in high school which prevented me from snowboarding in the first place. Mounted on the board going down the mountain was not the problem. It was when I had to unmount one leg (my good leg) to scoot me along (like a skateboarder) in the lift line. My bad knee, still in the binding mounted on the board, could not handle the twisted position of the leg (my goofy foot) and was prone to dislocate. I had no choice. I had to ski instead of snowboard. I was just thankful to be on the mountain!

After the Lacrosse accident I was fortunate enough to get away with orthoscopic surgery to treat the damage. Back then, my knee was probed and fixed without having to cut my leg open, and I didn't even need general anesthesia. I got to watch the seemingly simple procedure on a monitor. Not this time around. I needed full reconstruction the doctor told me which would require at least a couple hours under the knife. We set a date for December 21st, only three weeks away, during Winter Break..

It's amazing how your body heals itself. In a weeks' time the swelling went away, and I was walking normal. Had it not been for the doctor and X-rays I would never have guessed that my knee was missing a body part important to proper function.

The day before driving home for winter break, Angie convinced me to make a pit stop in Breckenridge. Visit a friend and enjoy a day of skiing. I was confident I could handle it because I was not experiencing any pain and I was walking and even running with ease. The swelling was gone too!

We were students. We couldn't afford a day pass at Breckenridge. We hiked through the trees and the deep snow to Tower Eight. From there we skied down to a lift we knew we could get away with hopping on without a pass. It was on mid mountain, so they didn't scan for passes at that lift. They assumed you had one because you can't get there without already having taken a lift from the base. *Unless you hike up!*

It was a great day. A blue bird. I skied the day without incident and enjoyed it to the max! A week and a half later, in the recovery room at Swedish Memorial Hospital, my doctor came in to check on me.

"I underestimated the extent of your injury," he said apologetically, "what I thought was going to be a two-three-hour surgery turned into a four-five-hour surgery. When we got in there, it was a lot worse than we expected. We couldn't use the tendon we planned on and we had to repair it as well."

"Oh really, wow, sorry, thank you, doctor." I didn't tell him about Breckenridge and how it was, no doubt, my fault. I must have put a lot of pressure on my tendon that day, which was obviously compensating for my missing ligament, and took a beating in doing so.

Ski or Bust

Good days, bad days
Blue birds and gray days
Snow days, slush days
High days and low days
A day on the mountain
Can come any day
Come what may
Yet always hoping...
Hope'n for powder days!

2020

Chapter 10 Postscript:

Three days ago, on Saturday, February 8, 2020, my brother, Solo, and I headed to Loveland Ski Area (which sits on top of Eisenhour Tunnel) to hit the powder! It had been snowing non-stop for the previous two to three days and there was going to be a couple feet of fresh snow. To make matters better, Loveland had closed the day before due to the storm and high winds so the snow would truly be fresh. A crowd would not have skied it off the day before.

"I don't like going up the mountain on the weekends," I told my brother as we were discussing it the night before. The traffic westbound on I-70 from Denver is a parking lot for 50 plus miles!

"It will be worth it," he replied. Not having to convince me, because I had already made up my mind. We were going. Ski or bust!

For the past 13 years I had been going to Loveland regularly. At first, I would buy a season ski pass, then when my brother (my other brother, Zeke) moved up to nearby Georgetown and began working at Loveland, he would always hook me up with a free day pass or corporate pass. Btw, my other brother, Solo, worked there for a season too, a couple years later. Nowadays, I purchase the Loveland 4-PACK because neither of my brothers work there anymore, and I just don't have the time or energy to go very often. Four times in one season is more than enough.

I love Loveland! I's only an hour's drive from Denver if you're not fighting traffic, and it's not over-crowded like Keystone or Vail or any of the other popular Colorado ski resorts. A lot less expensive too! A 4-pack costs $159.00 at pre-season sales; that amounts to $40.00/day which is unheard of in Colorado. Day passes at all other resorts are about a hundred dollars if not more! That's ridiculous. Lift way robbery if you ask me. The biggest complaint people have about Loveland is the wind. I concur.

One time, many years ago, the wind was so strong coming off the ridge it pushed our lift back so far we were almost parallel to the ground below us, having to hang on tight or risk sliding off. It was a tempest! A powerful gale would come over the ridge from the south, down and up the bowl to the north blowing upwards. The gusts would keep you suspended in mid run going downhill on a steep incline! The wind was so forceful that it gave gravity a run for its money! Skiing at Loveland, sometimes, is like undergoing arctic training. It's so intense that I make sure I only go on non-cold and hopefully non-high-wind days. It's located on the continental divide and so the strong winds coming over the ridge are cold cold cold! But I'm not complaining because it keeps the crowds at bay and leaves more fresh snow for me!

Speaking of fresh snow... three days ago at Loveland just about killed me! After fighting traffic for 3 ½ hours on I-70 we finally made it at 10:10

am. We had left the house before 6:30 am. The parking lot at the basin (upper mountain... where we wanted to be) was full so we were forced to park in the valley below and take a shuttle up. Bummer. Next time we will have to leave the house at 5:00 am!

After booting up and suiting up we made our way through the parking lot to the lift because we figured we would take a run in the valley since we were there. The "valley" was where you would find the ski school and all the snow-go-getters on training wheels. The bunny slope was full of tots and one or two grown-ups in training, snowplowing down, back and forth, in single row formation, with their instructor leading the way. At the bottom they would line up to enter into the "magic tunnel" my brother would say, amused. Poking fun. The magic tunnel was a covered conveyor belt (like the kind at airports) about a hundred or so feet long to carry the trainees back up the cottontail slope.

Adjacent to the magic tunnel was another lift. A low-speed quad. We shared the four-man lift with another, who was waiting his turn in the "singles line" to hitch a ride with the next group who had room to spare. The lift, itself, was identical to Lift One located at the basin, only it moved 100 times slower and climbed less than half the distance. The ride up was gruelingly slow. All we could see below us were groomed blue runs and already-tracked powder directly under our lift.

"Dang, we gotta get to the basin before everything is skied off," I said with disappointment clear in my voice.

"Yeah, what are we doing wasting our time here?" Solo replied.

When we finally got off the lift we veered right looking for a run that had fresh snow so we could make our own tracks. There wasn't any, but the snow was deep regardless. As soon as I dropped in (the ski run) my parabolic little boards sunk in the snow and it was too late for me to recover. I was stuck.

Literally. It felt like cement. I knew right then and there I had made a mistake coming down this way. 1. The slope wasn't steep enough for me to stay on top of the powder. 2. My skis weren't long enough either. 3. I wasn't strong enough to endure what was ahead of me.

I knew all of this as soon as I got stuck. This wasn't my first rodeo. During college I skied on 185's regularly and 190's on powder days. There were no parabolic boards back then. You needed longer ski's for higher performance. I had my 185 cm K2's, but on powder days I used my Head skis, at 190 centimeters I was hard pressed to sink in the snow. Hehe. *{If you haven't noticed by now, a "hehe" comes after a pun I find amusing, lol... actually, I went back and deleted many of the hehe's because it was getting obnoxious. Too many puns in this book... a literary dream come true}!

Today, my skis are 155 cm, but the widening shape to the tips of the skis, I'm told, makes up for the loss of length. Not today! I couldn't tell, anyway. The terrible part was that the skis were long enough to cut below two feet of snow and be cemented in. I had to dig out. Solo ended up hiking up from where he was to help me. I was able to get the back but not the front. Afterward, I had to rest to catch my breath, defog my goggles, tuck in my under armor, button my pants, put on "lippage" (my lips were dry), and suck on snow (cuz I was thirsty)!

"You got this, Gwyn, just lean back."

"I did!" Defending myself from my brother's criticism. Truth be told, I didn't have the stamina and endurance of my youth. I was weak in my body and esteem.

My brother was being very sympathetic because he, too, sunk and got stuck in his snowboard. I felt really bad for him. He was being so patient with me. He got stuck once, whereas I got stuck every 20 feet the whole way down. Halfway down the run, I was totally exhausted, and Solo was already at the bottom. Waiting for me. By this point the majority of my problem was that I didn't have enough energy to continue. Each time I was sunk, it took all my muscles to wiggle-wriggle and/or dig to come loose. Each time, I got weaker. Enervated. What little energy I had left was just enough for me to be able to ski down until I had sunk again! I was running on fumes. I just lied there on the snow, half my body sunk in cement, staring up at the patchy blue-gray sky. Tried to wriggle. Couldn't. I was languid by now and I decided to call for help. I couldn't go on. I needed rescue.

Unlike my college days, specifically the day at Irwin Lake on the snowmobile, I was not cussing, and I was not crying, but I was praying. Pleading in exhaustion. "Help me, Lord!"

After my prayer, I was able to cut loose. It was a miracle. It wasn't easy. It's not that my legs and skis suddenly became like a hot knife cutting through butter. It was not that kind of a miracle. I was able to muster the strength needed to get through it. Like it says in Isaiah 40, God gives strength to those who have none. He gives power to the weak. I was the weak. I was ready to throw in the towel and call for ski patrol to... not just dig me out, but give me a ride down the mountain. I was debilitated at that point. But, God answered my prayer. He restored my will and filled my tank just enough to fire up the pistons! The rest was up to me.

Long story short... too late, sorry, I didn't mean to drag this on... The last two days recovering was akin to having flu-symptom body aches. Even the muscles in my fingers throbbed, and I can still feel it aching as I type this now. I could barely walk. I could barely talk, even. My jaw ached. My toes. My head, but especially my arms and legs and *both knees*! To top it all off I slipped on the stairs at home and two days later tripped and fell on the floor and fell like a tree twice that week slipping on ice on my driveway. Talk about pouring salt on a wound! Oy vey! I feel a bruise coming on just thinking about it! Thinking back, not sure why February snow on the mountain was so heavy, like spring snow. Hmmm.

Getting up at 5am on a Saturday to fight heavy traffic on I-70 for nearly four hours...

It was so NOT worth it.

Non-local Vision

The haunted hole drilled by a holographic viper.
A hyper-
active interest in the hyperbole of hyperspace.
Trace
the face of the universe while caught in space that won the race with time.
Primordial
existence in discord by an accordion of dimensions.
Prevention
of inventions to create a mind to find the Holy Grail.
Mail
our muse to the moon and concentrate on fusion.
Illusion
of answers to the T-shirt equation...
Creation
of hope in a hopeless hunt for the Grand-Unified Theory.
>>>

{Continued from previous page}

<<<

Weary
to walk with ones who are leery of talk about what we don't know.
Slow
to admit our tracks may lead nowhere and an abyss is around you.
New
to the cosmic neighborhood caused by the colossal crash of atom.
Fathom
kinetics, inertia, entropy, and quantum altruism.
Magnetism,
electricity, both nuclear forces, and gravity.
Depravity
of density, the destiny of particles untamed.
Maimed
are we who believe to be one or three one must have eyes to see,
Free
are we who lack eyes but see raw simplicity.

1998

Chapter ΣLΣVΣN

Kiss

Math is a language. Languages are formulas too. Could you imagine translating everything we say and do into a mathematical language. First grunt. First gesture. First sign. First glyph. First words. First script. Simple enough for Maxwell to translate what Faraday was saying about one copper wire and one rotating bar magnet, but could you imagine if it took Faraday copper wire of varying gauge sizes and 14 magnets at varying rotations? So much for a T-shirt equation, right!

The Large Hadron Collider is the biggest machine in the world. Super complex. Have you seen pictures of that thing? The world wide web is probably the most complex machine in the world. Both come from the same place. CERN must have the world's largest whiteboards too! How would you like to translate that math! Oh, for crying out loud!

The internet has a lot of connected parts. It really is a web. But it's non-local to one place, yet every region or area has a place... a hub. You probably know that the information you send and receive in cyberspace is translated into a binary system of ones and zeros. Each one or zero is equivalent to one bit. There are eight bits to a byte. I sent one photo the other day from my phone to a computer in my house to print. The photo was 600 kilobytes (KB) which is 614,000 bytes, which is 4,912,000 bits. You would think because I was in the house, in fact standing in front of the computer, that the information (photo) I sent from my phone would go directly to the computer, especially because it got there so fast.

But nooooo... it's not that easy.

For that one photo, the individual bits had to travel on radio waves via wi-fi, then a pulse from the router to the receiver which is located at an internet hub and back again! I will repeat in detail so you fully understand: the data from the one photo, is broken down into four million nine hundred twelve thousand bits of ones and zeros, each one of them riding a separate wave on the wi-fi frequency "airplane" to the router and from there they have to hop off the plane and take a train. This train's not flying on wi-fi. Now the bits must ride pulsating electric currents on a copper wire or pulsating light from fiber optics to the receiver located at a local internet hub. Depending on where I live in relation to the nearest hub (which could be miles away), the bits must make the journey there. For one photo! All networks and internet service providers (ISP) go to the HUB because that's where the big receiver is... geez. The receiver is more than a digital post office. It is a translator. It is a sorter. It is a calculator. It is one helluva machine! It will detect the information using frequency modulation. Then it will repackage my one photo there and send it back out (via the assigned ISP) to the address on my phone through a web of wire pathways the same way they came in, each bit riding a pulse all the way back to my router then into my computer's inbox via ethernet. Amazing! Or, from phone to phone if I was out and about it would ride the frequency wave to a cell phone tower, making necessary pit stops to other cell phone towers (if I lived in the boonies) on the way to the hub. Either way, it must go to the hub. Makes my head spin! And you thought it all happened in an instant, didn't you? I did, until I researched it. *Don't get me started on 5G and LED's (not yet anyway). Coming just around the corner. For now, we are still dependent on wires and cables. Yes, even across the ocean. They do not bounce all this data from satellites across the seas. It is all connected with wire and/or fiber optics. Even the "cloud" is on hard drives. Big ones. There is no virtual puffy cloud in the sky holding all our data. Anyway, can you picture in your head the amount of numbers and symbols it would take to translate this entire process into the language of math? Never mind the gazillion bits of ones and zeros! I mean c'mon!

https://web.stanford.edu/class/msande91si/www-spr04/readings/week1/InternetWhitepaper.htm

Imagine now sweeping away all the cobwebs of the internet wires and cables, and radio frequencies, and taking away the router, cell phone towers, and internet hubs. Remove the computers and the ISP's and everything in between until you get to the bits. The slate is clean and all you have are ones and zeros.

Can you picture it? A lot simpler and more peaceful isn't it?

From Deal or No Deal to Survivor, from corporate takeovers to basic mathematics and scientific methods of probabilities, elimination processes are used as a function or technique to achieve a winner or a correct answer because it is a reliable way to a solution. Moreover, by eliminating the crud we uncover what is beneath; what is hidden. We can see the raw, naked, unadulterated form underneath. The processes to eliminate exteriors strips down coatings so we can see the nature of the beast. What is behind the mask? What's behind the wall? In order to find the most fundamental source of anything we must peel away the layers so we can see what's inside. The process of elimination takes us to the center of the lollipop... what's in there?

Everyone loves surprises. So do I and I will keep unwrapping as long as there is something to unwrap because I know the prize is in the center. The central component of life itself, holds the best prize of all.

Keep stripping, keep peeling, keep cutting away until you get to the base. The last layer or the last piece... the truth being the remains of what is left. Like refining gold, once you burn away the dross, you are left with purer gold. The purest gold is crystal clear! Clean.

Stripping down, scrubbing off, peeling away, or burning with fire.

You uncover something else... something deeper, something raw, something pure... it gets less complicated. Less dirty. The less you have the easier it is to clean, lol, and to polish. The less you have the less mess there will be. Eliminating layers gets closer to the heart of the machine, the heart of the matter. It gets closer to the heart, period. Eliminating all the fluff gets closer to the *truth*. The heart of the substance is the real treat.

I'm not talking about entropy this time. This is not the process of elimination as referred to in chapter eight. This kind is not harmful. It might be painful at times, but it's not going to contribute to death. No, this kind makes you thrive. This is the process of illumination. It's like cleaning off the dirt from a light bulb so it can shine brightly. It is to make things pure. The purest of anything will be free from ick. Gold or otherwise. If you don't clean off the ick yourself, then ironically, entropy will do it for ya! In cosmology, entropy is known as heat death. In mechanics it is known as unavailable energy in thermodynamic processes. We talked about this. What happens when things decay? Moldy bread, or browning apples? Weak nuclear force is ultimately at work which is fueled by the friction of its unstable atomic moving parts, the friction generates heat. The heat will eat. It will take bite after bite until it's all gone. Well, uh, unless it's pure honey. The ick has nothing to stick to... funny huh, honey can be icky if it's stuck on you, but pure honey is the only food that will not rot but it's oh so sticky! God has a great sense of humor!

Most people argue over what is pure and what is not, as if it's debatable. As if opinion matters. If we use Occam's Razor as the principle tool to determine bulls#*t from non-BS the lens from which we view purity will be clear. If you did not take the time in chapter three to look up Occam's Razor, here it is in a nutshell:

Keep **I**t **S**imple **S**tupid. No seriously, that's how I've always remembered it. K.I.S.S. KISS everything and everything will be much easier to understand.

Here's Wiki's definition, fyi:

> **Occam's razor** (also **Ockham's razor** or **Ocham's razor**: Latin: *novacula Occami*; or **law of parsimony**: Latin: *lex parsimoniae*) is the problem-solving principle that states that "Entities should not be multiplied without necessity."[1][2] The idea is attributed to English Franciscan friar William of Ockham (c. 1287–1347), a scholastic philosopher and theologian who used a preference for simplicity to defend the idea of divine miracles. It is sometimes paraphrased by a statement like "the simplest solution is most likely the right one". Occam's razor says that when presented with competing hypotheses that make the same predictions, one should select the solution with the fewest assumptions,[3] and it is not meant to be a way of choosing between hypotheses that make different predictions.
> https://en.wikipedia.org/wiki/Occam%27s_razor

Wise advice, no doubt. Just remember KISS. It's the simplest definition!

Now let's take the razor and cut into the cross. In chapter three we saw how the cross was the perfect symbol to unify the Ten Commandments. Here's a quick recap:

The Ten Commandments are laws on love. These laws can be divided into two parts: first half is God centered and the 2nd half is human focused. The perfect balance. Jesus said all the commandments can be fulfilled if we love God first and then love fellow man. Everything, He said, hinged on these two things. So, He said first we must Love God with everything that is in us. If we do that we can establish a foundational relationship vertically toward God. Heavenward, ↑. In addition, if we love the people around us we can establish a relationship horizontally with those around us. Side by side. Hand in hand. Outward, ↔. Once you combine the two you have the cross, † This symbol unifies the Old Testament laws! But let's not get caught up in a religious web.

The cross is not a symbol of man-made religion, as mankind so rudely would have us believe. This unified formula of the cross and what it represents is not some esoteric charm fit for only those few who find Jesus! The cross was no accident. The Old Testament predicted the HaMashiach (anointed one) would die on a cross 400 years before the cross of Roman capital punishment was invented. The Anointed One would be the light of the world and the son of God (YHVH).

Every number counts. And so does every symbol. Whether we add or take away. Let me ask a simple question. What happens when we cut the cross? It easily comes apart in two pieces, doesn't it? We have two pieces of wood? The one symbol of the cross has now become two symbols linear in nature. What do we have?

Don't over think it. *KISS.*

We have two straight lines | | and at this point it really doesn't matter which way you look at them — — vertically, horizontally, — | or one of each. For now, it does not matter. We have a line x 2.

From the earliest glyphs of ancient civilizations and in schools today the one-dimensional line is the most elementary symbol on the planet beside the dot. Any objections? I didn't think so. Keeping it simple. What does the single line represent? The number one? Sure. One finger, one line. That's obvious. Believe it or not, we can go simpler. The line represents a divide. A separation. It has a function. It is itself a knife. It cuts. It is the minus sign. We use it in math symbology to describe its function. To take away.

The glyph itself depicts its purpose. What about its nature? We know its function, so, now let's determine what it *is*... remember KISS... we have the most elementary symbol of all time! What is it? It IS TIME. Time cuts like a knife. Time, by nature, moves in a straight line (relatively). It moves from point a to point b, c, d, e, f, g, h, I, j ∞ from 1pm to 2pm to 3pm into eternity! The line is time. It represents the single dimension we know of as TIME. Time is a line, and it is the atomic knife.

Keeping it simple, let's continue. We have two pieces of wood. What does that mean? I don't know about you, but when I was a kid, I was taught that in the *olden days* two pieces of wood was all they needed to make fire. Fire sticks. It is burned into my memory, and I'm not talking about an Amazon thumb-drive you plug into your TV! I picture a prehistoric setting with a guy who looks just like the Geico Caveman sitting in front of his cave rubbing two sticks together ferociously. Two sticks make fire. I know better now; it's not that easy. It can be done, but exhaustion will settle in before you achieve enough friction to form an ember. *If you want to try this at home I recommend the bow drill technique with lots of dry tinder! Lol.*

As difficult as it may be to make fire by rubbing two sticks together, the concept is legitimate and conclusive... it is another elementary function. Friction makes fire. But what is its nature? What do you see? Other than primitive heat? I see a spark. What is the spark? Energy? Light?

The Quantum Cross

It was a sunny yet chilly afternoon in Bozeman Montana that day in early October 1997. It was my day off from work and a co-worker came over to play a game of Chess with me, which was pretty routine back then. I had been living alone trying to cope with the break-up between me and my ex fiancé, Caleb, just two months prior. Yeah, we ended up getting engaged after I graduated from Western, but not before we moved to Bozeman. He had always wanted to climb and ski Big Sky Country, and I wanted to attend graduate school at MSU. We packed up all our belongings and threw it in the back of Caleb's Ford F150 pick-up, made a pit stop in Denver then headed to Montana driving north on I-25, then west on I-90. We drove through the night. I remember waking up in the morning to gorgeous green mountains hugging us on both sides as we were making our final stretch from Livingston to Bozeman. Compared to where we came from the mountains here were noticeably greener. Lots of ferns; Colorado doesn't really have ferns adorning their forests. I didn't realize how brown the Colorado Rockies were until I noticed how green the Montana Rockies were!

Upon arriving to Bozeman, we began looking for a place to rent while we camped for over a month in Hyalite Canyon before we found a nice single-family ranch to call home. Long story short: a couple years later we broke up. I was grieving and to make matters worse, there were "bumps in the night" keeping me from sound sleep. I don't know if I was haunted by spirits or demons or a grumpy old couple known as the boiler and furnace, but I was grateful to my supervisor who took me under her wing with love and care. I worked and kept busy most of the time, picking up some graveyard shifts. A few of us from work started getting together to play chess.

The home Caleb and I shared was still haunting me just days before my friend and coworker was coming over for our regular chess match. My supervisor's relative, a pastor's wife from Helena, graciously came to my house to pray with me and cleansed it. It worked, praise the Lord! I had turned over a new leaf and was finally free from the sleepless nights and the dark days. Now my home was fit to host friends and games.

We sat down and began to place our chess pieces on the board accordingly. Sliding one of my pawns two spaces forward was my first move. The instant I removed my hand from the pawn I had what many would consider to be a vision. For me, it was much more than an out-of-body experience. It was an "inner" body experience. Have you ever seen the movie, Honey I Shrunk the Kids? It was as if the fabric of space, itself, ripped open and sucked me in as it *shrunk me through* to the other side. Instantly, I could see the sub-atomic world around me. Within me? In that instant I could see particles of light and the attraction it had to other particles. Simultaneously, a wealth of information pertaining to it was downloaded into my soul. Some might say I had an "a-ha" encounter and minimize it in so doing. I had had moments of enlightenment and this was NOT THAT. This was not a wet moment either. And this was more than an experience of transcendental awareness from an acid trip. Let me tell you, I am almost 50 years old and I had never before experienced anything like it and since then I had only one more experience (similar but smaller). Fyi, I continue to have visions and dreams, but not like this!

As soon as I let go of the pawn, and the "vision" began, everything outside of this vision, this interdimensional window, became a blur. I do not recall my friend leaving my house, but I do know that at that moment it was game over. It stopped just as soon as we began. I remember saying, *can you see that?* But that was the extent of our dialogue. It is so hard to explain what I saw, but I can tell you this: I, instinctively, knew that no two things can occupy the same space at the same time. Nothing. Like pieces on a chess board. The queen will take the space of a pawn, thereby killing the pawn. Like you and a Mack truck. The truck will win. Like me and the snowboarder, my knee took the "kill" and my ligament died. I was understanding this truth starting with tiny particles. I didn't know what I was looking at, but I fully understood what I was seeing. I saw atoms, but I didn't realize it at the time.

The atoms were unable to crash into each other naturally. On a molecular scale, I saw cells die when their "space" was occupied by an "invader." Viruses? Bacteria or some other parasite? In my minds' eye I even saw the Mack truck scenario! I also saw a knife plunge into a man and kill him, but then saw how the knife entered the place in his gut which had no room to share space with it and his organ couldn't be repaired in time. Then I would be able to see what happened in the gut on the molecular level, and then I was back in the atomic world to see how it affected him there, eventually as he decayed and turned back into ash. No two things can occupy the same space at the same time. It is lethal when it does.

The event lasted three days, and in those three days I did not eat, nor did I sleep. Between visions and heavenly downloads, I cried out to God, praised Him, journaled, and wept some more. Mostly tears of unimaginable joy and peace and love as God cradled me through the entire event. This was a landmark occasion in my life. Looking back, it was a pinnacle milestone that changed me both inside and out. I came away from it with a deep understanding of how two things cannot occupy the same space at the same time, *except for one thing*. I also came away with some interesting formulas. The following two years can be summed up in three words. Book worm hole. *Lol! I became a bookworm exploring wormholes.*

The Bozeman Public Library was my home away from home when I wasn't working. I was only allowed to check out four books at a time, so I spent much time there researching. The Internet was coming out of its infancy and it didn't have all the answers, like it does today (my eyes are rolling, lol)! I must say, the content was more raw back then, and it was easier to find good material. These days, as numerous as the options and answers are to our Google searches, it has been scrubbed and diluted. Sad truth. The internet today leans more to quantity over quality, and much of the quantitative data is manipulated by you know who! You just can't trust it like you used to!

I had a new computer and I loved it! It had a three-gigabyte hard drive and four megs of ram. Wow. Thinking back, it was amazing how much I could do, and did do, on that system. I think I still have it somewhere in my basement. Hoping someday, miraculously, I will retrieve the stories I wrote when I first started writing before it crashed. Lol!

After the three-day event of my atomic-out-of-body-experience I was driven to research to understand the basic equations I had written. Well, uh, I DID understand what I wrote but I didn't fully understand what I understood, and I didn't know if others would! Does that make sense? I was not a math person by any stretch back then, philosophically or otherwise. Although, I enjoyed Algebra and amazingly made A's in it in high school; that was the sum of my knowledge of math. Oh, I did love algorithms! I did well enough to test out of math in the college competency exam pre-freshman year, so I was excused to never take math again. And that is exactly what I did! In fact, Angie and I threw a "never-have-to-take-math-again" party in my dorm room the first week of school!

Here I was with some formulas I had written down and I hadn't the faintest idea what to do with them. They changed the way I viewed the world and I wanted to know more about them. I had to find whether someone else had these same findings. Could this summation be for real, I wondered? With that in mind I began a rigorous and wholly devoted quest to have understanding because my body was buzzing to share them. I needed to learn the language. I was utterly clueless. But not for long.

It was as if God turned on my switch! He pushed a button which supernaturally allowed me to learn and understand things I had previously not even had a desire to learn. For, example, I would watch someone play Fur Elise by Beethoven on the piano at work and in less than a week I would be able to play it myself! With two hands! Whaaat! I'm not exaggerating! This went on for years! Before that, the only thing I could play with both hands was Chopsticks! They say practice makes perfect, so I must admit, I am sorely out of practice, and cannot remember most of the songs I once knew how to play. What's more, I would devour 300 plus page books in a weekend. Books about math! Books about physics! Books about theories and philosophies of time and space. The concept was so strange to me, but I found it to be another miracle that the first three books I checked out... not only could I understand what it was saying, but it applied to my own discovery to a tee! *I have a grand unified theory! Holy cow! I think I have THE Grand Unified Equation! Stephen Hawking would agree! My formula is smaller than all the one's he threw away!* I kept going over it in my head. I had the ultimate T-shirt equation. It was sobering. I couldn't find any book that already laid claim to it. I did read about those who claimed to have it, but it wasn't the same as mine. String Theory claimed to have it, but there were so many holes in that string! I knew I had it because, indeed my formula made sense, and indeed it was smaller than all of *theirs*! But I had to be sure.

I ravaged every book at the library pertaining to atoms and the subatomic world. These led me to books on astrophysics. The subatomic world is not too different from the astronomical! I learned about quantum mechanics and calculus, and I read then-current literature and columns from publications regarding recent findings and new theories about parallel universes, hyper-dimensions, and energy. Oh, I especially loved learning about the origin of numbers. The origin of anything and everything. I was captivated. For the first time in my life I had a purpose! I looked but could not find a graduate program for math philosophy, I just found plain old math itself. *If I didn't make it clear before I will say it again for the record*; I do not claim to be a mathematician. I can barely do basic math in my head. What I do have is a *passion for math*... the poetry of it, the origin and philosophy of it. The language of it. The way it is used as shorthand to explain functions intrigues me. If I do not understand the translation of an equation I move on. As much as I love and can appreciate the complexities of numbers and formulas, convoluted equations are elusive to me. Too many variables and factors make my head spin. I prefer T-shirt formulas! I will leave the complicated math to the professionals.

I thought how funny it would be to be a Doctor of Philosophy in the *Philosophy* of math. But a PhD in math philosophy didn't exist. Not that I could find, anyway. I chuckled, sighed, and I shrugged it off and refocused on my own writing. It was all I had, and I was running with it! Unless someone beat me to it I was going to introduce to the world the Grand Unified Theory to trump all others. I withdrew from all attempts to go to graduate school, instead, I will write this book, I told myself... oh, and ski and mountain bike and climb mountains! Lol!

Although I understood the contents of these "far out" books, I still had to turn to other resources for definitions of terms unfamiliar to me which led me to textbooks which led me to the history of it all. Plato and Aristotle, Copernicus and Newton, and of course, Michael Faraday! Then there's the more recent scientists, such as Heisenberg and Schrodinger and their amazing discoveries in the quantum field. And of course, Tesla! Just to name a few. There were/are so many! As I read about all their incredible discoveries, I had so many questions I wanted to ask them. I couldn't believe I found myself in a place where I felt comfortable to have a conversation with them, "if they were here." It hit me all at once that I *now knew* the language. I knew enough. If I had the honor to meet with any one of them I could converse with them confidently. I was astounded. This fueled me even more, and I kept returning to the library. I recall after that first visit, when I went to return my first three books I had a dozen more books in my arms before the library was getting ready to close. As I made my way to the check-out desk, the librarian looked at me and shook her head.

"You can only check out four books at a time," she said with a smirk. I remember her fidgeting while I had to decide which four to take home.

I loved learning about the greats, and the underdogs too. Faraday will always be among the greats to me, even though he was considered an underdog. During this time, I became especially infatuated by Albert Einstein, his life, his discoveries, his formulas.

Einstein arrived at his Special Relativity and onto his General Relativity from first discovering what he called the Shrinking Factor, he combined it with Kinetic Energy (a T-shirt equation). If you know basic algebra, you can easily solve this one yourself. The Shrinking Factor is a T-shirt equation too.

$SF = 1 - (v^2/c^2)^{1/2}$ plus $KE = M \times 1/2v^2$ ... this led to $E=mc^2$. What a lovely combination. Very harmonious. *Unnecessary tangent, sorry. Where was I? Oh yes, my vision.

No two things can occupy the same space at the same time except for one thing: Time. I repeat for emphasis, no two things can occupy the same space at the same time except for time itself. Unto itself. The two sticks making fire is representative of that. *Whether there are multiple streams of time intersecting time, I don't know. I only saw one in my vision. One intersecting set, like two sticks. Each stick representing a line. The line representing time. It wasn't until years later that I saw time moving (perceivably in a straight line) in a gradual curve encircling a universal expanse and eventually intersect itself. Like a figure eight. No wait, I saw a

Jesus fish without the pointy nose, lol. The point here is I saw the intersection. For the sake of explaining my original vision and staying on course I will stick to that *line of thought* as we continue. *Hehe.* Two lines. One line, squared.

Time cuts through space like a knife. Time will eventually cut through *our space* and leave us in ash. Time times itself equals the cross. *In pre-math language, minus sign squared equals plus sign.* Lol!

$$-^2 = +$$

What is the plus sign?

Look at it. KISS.

It is space. One-dimension squared equals two-dimensional space. It is the culmination of space itself in its most rudimentary form... it is a prefatory explanation of the glyph. Space in relation to time has more than one dimension. Time is linear whereas space is not. It reveals the framework from which we calculate spatial dimensions, just like in the Cartesian Coordinates. The intersection of time causes a spark. That spark is a result of friction. This joint is cold fusion in the raw. Time by itself is cold. It has no way of generating heat because it is linear. It is as *one stick*, and you need two to make fire! Once the connection is made the (cold) two become (a hot) one. I saw this in my vision. I saw how time intersected itself. It was the spark of life. On a grand scale cosmically. And on a subatomic scale. This is how light is ignited. Two sticks. Voila! Fire. KISS. There's more.

$+^2 = $ X Now multiplication is born!

Time, by nature, is in motion... it is the *verb* of life, whereas space is stationary... it is the *noun* of life. Although space is stationary it is alive with movement. As we learned in chapter two everything moves.

Everything vibrates with energy. The *movement* is time. You can't stop it. Time is unstoppable. Once it intersects itself it is still in motion but now it is anchored. Like a wild horse in a corral running in circles.

Now that we have the cross, the *axes*, what comes next? Or rather, what came with it? You know this one.

Zero.

Time is still in motion but now it has been harnessed. It's still moving but no longer running in a straight line. The river of time has now been dammed up and its energy is contained. Time. Time is the knife that cuts through space, but it is also what creates it. $Time^2$= space. Time gives us space when it intersects itself. Time is still moving but is now moving in place. Living space. It opens a "reservoir." Time makes space. Space in which to grow. + is two-dimensional space and is represented as the point of friction. Fusion. And x represents the growth as it moves into three dimensions as it multiplies.

If − is time and + is space, what's o? A birthing canal? A gate? A continuum? Of the space-time-*continuum*?

Zero. The gate that opens and allows the numbers to pass. Like all gates, they open and close. The place holder that it is, it can hold in place the unstoppable vortex. The zero can close the hole as it retracts. Zero is a gate. Zero is a valve. Zero is a womb. Remember at the end of chapter nine, I said the key was in zero? The key is the cross and it literally is in the zero. It *is the key* which unlocks the gate. It controls the gate. It is the tether that keeps time from cutting into everything and creating a bloody mess! Time is the wild horse of which God tamed, and the cross is the bridle.

Time. Now it is captured and put in a jar! Like a lamp. It creates heat. It is the source of heat. It is the light and it is the source of light in our material world. It is alive. It fashions the womb. It is where procreation begins. Where multiplication takes place. It is the birthplace of all birthplaces.

This − is male and this + is female.

÷ What do you see? The division symbol, right? In light of everything what else do you see?

Time and space in relation to one another. Just like the knife that it is, time cuts through space dividing it in half. In all my research I couldn't find a satisfactory origin of these elementary math symbols but the minus and plus signs were very predominant in Chinese history. Also, Sumerian stone tablets showed numbering systems with a minus sign in their calculations, but their minus sign looked like a boomerang; two lines joined at the point of a 90° angle. The Mycenaean Linear B Numeric System contains lines and circles beginning with their number one as a line and 100 as a circle (see chapter six). Ancient Greece had a vertical line with a dot next to it and used two lines to contain dots. The Brahmi of India used a plus sign for its number four. I wanted to continue my education to learn more about ancient symbols. Surely, the true origin of these symbols must come from the naval of earth, I told myself! It is indicative to space and time. It is no accident.

The sun cross ⊕ is one of the oldest symbols I found but its recorded origin is not local to one religion or region. It appears in ancient glyphs around the world. In my next book we will learn how everything seems to funnel to the naval of the earth, for now we will continue with time and space and how it is quantifiable in stacking the Jenga blocks of life.

What an incredible coincidence that the symbols we use to determine the fundamental calculations of numerals are in of themselves accountable in the counting! And that the glyphs are explanatory of their functions! Especially with time and space. It gives new meaning to space-time continuum you can picture in your head! It no longer is an abstract image or an idea only for scholars. It's no longer a muddy concept, right? It is plain as day. The atomic knife is time itself! The cross is the math maker. The number bearer. Just as we saw in chapter nine with the positive and negative vertices of the parabola, each image of the graph supporting the function in of itself exhibited the fact that the function could not be defined without the points and axes. In order to find a unified explanation of any function, including life itself, we have to look at the *forest beyond the trees*. By and large, we are too concerned with the functions of life we overlook what's most obvious. The *ruler* of the function. The graph maker. The coordinate maker. What is the scaffolding holding the builder if not the framework to create the framework?

All the outspoken physicists in all the books I've read insisted that the Holy Grail of Physics must be simple and unadulterated. Must be very small. Small and simple because it's what's left after you strip everything away; Everyone is searching for the fabric of space and talking about string theory, but overlooks the needle pulling the thread! I'm told the grand unified field formula must be elementary in nature and meet the requirement of a true singularity. Doesn't it make sense, then, to achieve such a grand equation we must enter a pre-numeral state? How much smaller can we get? The students that were presenting their unification formulas to Hawking 30 years ago were using numbers in their math. Regardless of how many or few numbers they used in their equations, using even one number is too big. Even number nine. Especially number nine.

$-$ $+$ X $\div$ each have a function for space and/or time. Even the equal sign, like two lines of time pointing in the direction where they will ignite and give birth to another!

$$-^2 = + \quad +^2 = X \text{ and } X^2 = O$$

In my vision, zero appears once the "spin" begins which is signified by the + turning into x only as it rotates and multiplies from a single point. So, it could also be true and written as X = 0.

Zero opens into a womb. It is essential to giving birth... to multiplication, developing into numbers (countable particles). It just grows from there. You cannot have the zero (o) without the cross in motion. The zero is born out of the point of intersection as it whirls like a lasso. It came from the point of "spatial nothingness" except for time itself and widens to form a protective cell. Or a lightbulb! Time makes space. It pulsates and can widen and shrink as it spins conically. The length of the lasso (the arms of the cross) contributes to its magnetic force. It can attract or repel depending on the shape and force and direction of its rotation and/or pulsation. Spinning outward or inward will regulate its repulsion or propulsion.

The middle of the X. In math there is a big difference between + and x as it defines the functions of the numbers to which it is applied, but as symbols depicting time and space it seems similar. What's the difference? The x is the + in motion. When time occupies itself, it concentrates its energy into a single point; the instant (the Planck instant) that happens a new movement begins. This is stationary: + and this is not: x. The cross depicts the pre-quantum singularity, whereas the x depicts its quantum function once the spark is made. Time no longer moves in a straight line once the joint is fused. It's spinning and naturally creating a centrifugal force because time functions in constant motion and now has no choice but to move in rotation like a plasma ball. But, instead of a ball it forms a ring. A zero. A gate. And a womb. This is what I saw in my vision. Can you picture it? I know you can! If it weren't for the cross the simple illustration you are materializing in your mind's eye would not be possible! Now picture this: without the cross the zero would spin out of control. The cross is the anchor. The tether. It gives it gravity. Balance. It pulls it back in like a dog on a leash. Without the cross we would be blown into oblivion! The energy from "time gone wild" must be reeled back in.

Although I didn't see it in my vision, I imagine the opposite is also true. The momentum can move centripetally (a pull from the central joint), strictly to start flowing the other way. Into itself. A vacuum. A vortex. A black hole. Everything gets sucked back in. It is unstoppable. Space crushed back into linear time. Death in the physical happens. Unless you can stop it, it will spread. Correction: Time does not *spread*, but the space and matter being pulled back into it gathers and concentrates in its bottleneck... as it grows (like a traffic jam) it *spreads*. It is like cancer. Perhaps it is cancer. It cancels life. I bet 9 is responsible!

The singularity. The single point of space in time, and vice versa; the single point of time in space. It is where time stops and where space begins. Life is created. Life and death. If the cross is torn, time will move within that single point and resume its momentum in linear fashion into eternity, cutting like a knife all the material in its way! Times two! Times two x2x2x2x2x infinitum until there's nothing left. Even the ash will blow in the wind. Time moves into eternity. Space is only as good as the light that creates it. When the light goes out so does *its space*. The cross represents both life and death. It is the ultimate paradox. God has an amazing sense of humor and wit about Him. In chapter three, we saw how the cross was a sword. A sword cuts. It represents death. We also saw how the cross was seen in the cell-adhesion molecule, Laminin, which represents life and connectivity. Cross= space= time times time.

+... is this right side up or does it matter? If you turn the + to get a X the single point in the center is still there. Does it matter which way you look at it? Yes and no. It doesn't regarding its position in of itself. It does in regard to our position. Jesus was crucified on a cross, not an X. As you know, I do not believe in coincidence as random and meaningless. When two things coincide, it means something important just happened at the same place at the same time. Incidentally, the definition of coincide means to *occupy the same place in space*. Remember that the next time you run into a coincidence. Take notice. Anyway, Jesus died on a cross and not an X because God was giving us clues. In chapter three we saw how the cross unified the old covenant commandments and pointed to relationships with us and God, as well as with us and those around us. Without the vertical piece, the horizontal piece has nothing to support it. God was showing us that once we establish a foundation with Him, rooted and stable, we can successfully have meaningful relationships with fellow man into eternity. Forever and ever. But there's more. With God, there's always more.

+ ... I also trust God's little details. The Cross, unlike the X, symbolizes balance founded on the vertical line. The X symbolizes balance *equal* to each line, therefore the meaning is lost. God gave us a formula to first put him first, and to second put others second... to Him. He gave us this formula for our own good and for the good of all our relationships. When we are founded in Him we are better for it. *Once we took God out of the schools, all relationships fell apart. Man and wife. Kids and kids. You know it! Whether you believe it or not does not matter. We were all affected.

Gravity, I know, had been a struggle for Einstein til the day he died, and in particle physics it is still missing from the standard model. It is the thorn in their (physicists) side. And today, I have yet to read a satisfying unified explanation of it. And until today, as I sit here at my computer, I didn't have a satisfactory view of it either. But, as I typed the last two pages, the Holy Spirit came upon me and revealed the cross to me, again. And again. I returned to this chapter each time to add and omit.

I always knew it had to do with the cross, and Jesus was the key, but formulating it was a difficult task because God didn't show me. The original vision was missing gravity. But now, all is revealed to me thanks be to the King of Kings. The Single One who made all things. The alpha and omega, the beginning and the end. The straight and narrow, the circle and the bend. The Great I am who became the least. The least of these to slay the beast. The Life of Love and the Love of Life, His blood spilled to remove all strife. The magnanimous mystery of Life and death is found in the cross from His last breath. His last breath gives breath to me. To live and breathe eternally!

Now where was I? *I had to blow my nose and dry my eyes. ☺

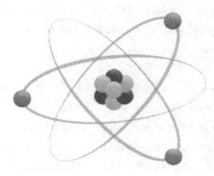

51. Atom, Orbital Model

Do you recall from chapter three about the standard model? The nucleus of an atom contains protons and neutrons, and the nucleus is surrounded by electrons as they orbit the nucleus. The electrons are considered indivisible (remember Democritus' atom) but protons and neutrons contain even smaller particles.

Remember this chart from chapter three (figure 52, next page)? This standard model of indivisible particles is what we will find in the nucleus, as well as in the perimeter "storm cloud" of the atom... its atmosphere. The photons bounce back and forth from the body of the nucleus with atmospheric electrons. Like lightning! Positive and negative ions igniting. Same difference, essentially.

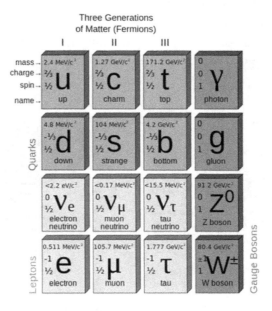

52. Standard Model of Elementary Particles (older version)

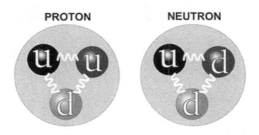

53. Protons have two Up Quarks and one Down Quark giving it a positive charge, and Neutrons have one Up Quark and two Down Quarks and remain neutral

Figure 53 is a rudimentary diagram because according to the experts, who smash these particles together with their huge toy, the protons and neutrons look more like a hodge podge of Ups and Downs, Antiquarks and Gluons (see figure 54). The subatomic packets with their varying charges combined in total determines the overall charge of the protons and neutrons within the nucleus.

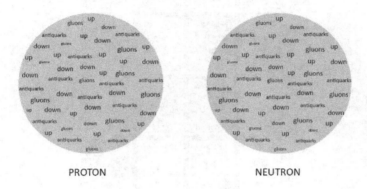

PROTON   NEUTRON

54. Protons and Neutrons

At first, only Quarks (the Up and Down ones) were thought to be the fundamental building blocks of matter, then they found more quarks: Charm and Strange, Top and Bottom... and they are paired that way, just like Up and Down, one with a positive charge and the other with a negative charge. Their total sum of these particles determines their positive or negative charge.

The negatively charged electron, to be more specific, has a brother and a sister which make up the family of Leptons. They are basically heavier electrons. Oh, and let's not discount their neutrino (neutral charge) counterparts. Ghostly doppelgangers? Lol! For the record, Tau and Muon have not been classified as male or female, lol, so they could be brothers, sisters, dad, uncles, etc. I'm just having fun explaining the family of particles. *The Quarks are quirky aunts and uncles, btw. Charm, Strange... Lol!

And then you have the Bosons (they're the Jedi's). The Gauge Bosons make up the indivisible particles which carry the "force." The Gluons carry the Strong force and the Photons carry the Electromagnetic force. Then there are the *bozos*, lol. The W and Z Bosons carry the Weak force. These "weak" bosons allow protons and neutrons to transpose into one another via nuclear fusion and cause leptons to decay. Neutrinos are also governed by bosons as they can only interact with other particles via the boson's weak force. Are all in the family? Entering stage left is the long-awaited master jedi.

The Higgs Boson, which is not on the chart above (see figure 55), was a theoretical particle when I first read about it in the 90's. If you recall, I mentioned the Higgs Boson in chapter two, and in chapter three I shared what Dr. Lederman said in his book, The God Particle; that the Higgs Boson was sought after because physicists believed it would reveal the ultimate T-shirt equation and indeed *be the God Particle*. CERN found it and says it behaves as a quantum ripple in the field of energy. They say that the Higgs is responsible for giving mass to the other particles because its field (Higgs Field) allows the other particles to randomly collect mass as it moves through it. Like a snowball. They admit this is a theoretical explanation and they use molasses to describe it (see https://youtu.be/tcHz3o4t6Rk). This discovery in 2012 using the LHC (Large Hadron Collider) at CERN, which earned Peter Higgs and Francois Englert the Nobel Prize in Physics, has changed the standard model and the model is still developing as colliders uncover new particles yet to be discovered. It gets even more complex. Hadrons which make up baryons and mesons are particles that are not on the chart above or below. Hadrons contain quarks but is subdivided by mesons and baryons, then there are the newly discovered kaons which is a part of the meson family (a third cousin twice removed)? I dunno. To even make a comment about these subatomic particles I would need to go back to school. Or back to the Bozeman Library, lol!

Standard Model of Elementary Particles

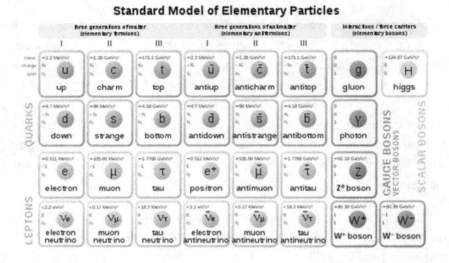

55. *Newer* Standard Model of Elementary Particles

You may have noticed that the chart (new and old) has a header to the left that says Three Generations of Matter (Fermions). Quarks and Leptons are collectively Fermions which spin at half an odd integer (1/2, 3/2, 5/2) as opposed to the whole spin of bosons. The Bosons are whole and are carriers of the forces. Why there are three generations of Fermions, or three pairs of each remains a mystery and is still being investigated. It has been determined that all these particles have antimatter duplicates. Antiquarks. Antileptons. You can think of them as their twin opposite, and they cancel each other out when they collide. So, if you ever meet your "doppelganger" twin, don't touch him or her because if you make contact you will annihilate each other. Lol! Then, you have dark matter. This stuff is supposed to be everything we cannot see or detect, so antimatter may fall in that category. Yet, dark matter is not the same as antimatter. It is surmised that most of the universe is made up of dark matter. In theory it is the stuff inside a black hole, and it is the stuff that is between all the stuff. As I said, dark matter is everything the professionals cannot detect. Empty space. Space without time? Time without space? Oy vey!

How do I reconcile what I saw in my vision with what we know about particle physics? I know what God showed me; therefore, I know in my heart it must be true. To take what I saw to coalesce with what they found may not be a task for me alone. Yet, despite new discoveries made in particle physics I do not see anything that would contradict what God showed me in a vision over 20 years ago. I would like to offer this elementary explanation for elementary particles as elementary truth:

We know a photon is a quantum of electromagnetic (EM) radiation, which is the smallest quantity of radiant energy. The photon particles are exchanged by the orbiting electrons and the nucleus by EM force. Photons have a positive charge and electrons (all of them except the neutrinos) have a negative charge. The family of Quarks also have positively and negatively charged pairs. These are all indivisible particles (Democritus' atoms). Do you recall them? The standard model of atoms imparts three out of the four standard forces. Electromagnetism and strong and weak nuclear forces. Gravity is still missing from their model. I would endeavor to say that gravity is not outdone by any of the other three forces. I'm certain, gravity is central to the other three. Literally.

The cross unifies them all. Without the cross the energy of time moves, in force, directionally contracted as a line. The cross harnesses it and tames its direction centrally, conserving the energy from a single point and allowing it to move within the circuit of the cross. As aforementioned, this new motion causes propulsion or repulsion. I postulate that propulsion gives a positive charge and repulsion gives a negative charge. The interactions between charged particles are, therefore, governed by the electromagnetic force. The cessation of propulsion or repulsion cancels any charge and makes it whole like the bosons. The Photon shines brightest as repulsion or propulsion gusts are not flickering the "flame" of its light and carries electromagnetic force the same way earth does. Earth, *as a positive ion*, interacts with the atmosphere of many negative ions. Photons appear to be the "lightning" between the nucleus and the outer cloud of electrons. Other than radio waves being ignored as a weak proponent of the Weak force (more on this later), I would submit that the weak and strong forces are adequately explained already from what is known about the gauge bosons. Yet, I would venture to say that the gluon which carries the strong force, is the "rebar" at the atomic level just as Laminin is the rebar on the molecular level. Gluon, like laminin, would have to be unlocked for us to *walk through trees*, lol. It is part of the circuit! The soul and spirit when divided by the sword of God will separate flesh from spirit beginning at the atomic level, would it not?

Then you have gravity. So perplexing. The key to solving this enigma is in the cross. As I tried to illustrate the importance of the cross being right side up + as opposed to diagonal x, I hope you understood the relevance. Not only was God declaring to us the priority of keeping Him first in order to maintain harmony with others, but also He was exposing the balance beam. The cross as opposed to the x explains gravity beautifully. Balance rests with and in gravity. The scales cannot be weighed without it. The upright cross depicts this to a T. Jesus, himself is the horizontal beam unified with God as the vertical beam. The Father's grounded vertical stand gives us the pull of gravity while the balance of the horizontal beam proves of its existence. In the same way Jesus proves God. We know we cannot have balance without gravity, but gravity is not detected without balance. The arms of the cross provide an "equatorial beam" centralizing the force. The whole thing is The Body. The substance with the weight. You cannot have gravity without a centered body. It is central to gravity. Literally. An uncentered body shifts gravity to one side or the other, which can be seen in the balance of the beam. The beam is the gauge. It detects the gravity. If gravity is off centered the beam will expose it as it leans toward gravity. Toward the weightier end. If it's level then gravity is centered. If not, the beam will disclose it. But the beam helps to support it and maintain its balance as it controls the equilibrium of the poles. God is the vertical beam giving us the poles. Jesus is the horizontal beam balancing the force.

Gravity, whether in a planetary body or a particle, if centered to its core will be proved so by its *beam of balance*. Both the vertical and the horizontal beams and every "beam of light" in between pull the force to its center and makes weight of it. You cannot prove gravity directly. You prove balance first, which proves centricity which proves gravity. Centricity determines equilibrium, and therefore, since we all, in Earth, are not walking sideways it proves Earth's gravity is centered. The cross is the missing piece which eluded Einstein, and it is the piece eluding physicists today. Without the cross our functional existence would be oblique. Our "uprightness" proves the earth's beam is in balance. Our personal "un*upright*eousness" proves our spiritual beam could use some adjustments and reparations.

The EM force. The Strong force. The Weak force. And the Gravitational force. It's all the same force deposited in different ways. It's one unified force. Energy is energy. Clean energy or dirty energy is determined by the technique which man uses to transform it and the exhaust it leaves behind. The energy itself is pure and simple. It is God given and we should thank Him every day for it. Why does man muddy the waters and turn everything into a complex complex? They can't see the forest for the trees because the web of which they wove and in which they move is tangled in a zillion numbers.

The EM force is governed by its polar charge (the propulsion and repulsion). The strong is governed by combustion, as fission (a crack of the whip) and held by fusion. The weak is governed by excretion (a leak). And gravity is governed by the centric equilibrium of the force. It's the Tootsie Pop surprise! I would bet that at the center of every "body" is a cross. A light and a womb. Could earth's outer womb be the abysmal ocean of our seas protecting the burning cross inside the center of earth's lollipop?

Micro-bodies and macro-bodies. In the quantum field and in the astral field. Oh, and in the Higgs field. This simple system of time and space is seen in everything. After my vision, I saw it everywhere. It's massive. It's miniscule. It's all over the place! It is not some hidden secret. It was hiding in plain sight. It is the proverbial writing on the wall, the symbols on every chalk board and white board and caveman wall since time immemorial. If one had eyes to see. Now that I've presented my simple equation, and painted a picture to explain it, I know you can see it too! KISS!

When time intersected itself, it became *The Singularity* via cold fusion. The singularity was the ignition point and from there it was able to concentrate its energy to its core creating a womb ((zero) in itself) which, in turn, was able to culture the energy into a bubble of space or a baby particle. The expansion of that space pulsates as it grows like a twinkling star; could this be the frequency of the quantum ripple of the Higgs Boson? The space is encapsulated by zero itself as it incubates the energy woven by the gravity of God as it grows from the center outward. It makes sense. It's easy to picture. Now contrast this with the counterintuitive explanation offered by CERN. Consider the theoretical Higgs Field which claims life grows as the Higgs particle picks up random particles as it moves through a "molasses-like" field. I call it their snowball effect. A snowball gains mass from its exterior. I would offer to replace their snowball with my particle baby. Their snowball could be the X from $+^2$ as it multiplies from the womb, right? Not a snowball from their molasses metaphor, but a baby from my simile: the zero giving birth and rebirth with every spark of the cross from the inside out. After all, exterior attachments result in ick and decay, I recall. Snowballs grow by picking up particles from its exterior. Nothing in life grows from the outward in... but icky stuff does! Life is not a snowball! Cancer is! It makes more sense to grow from the interior out than the exterior in… just like life shows us in nature. Sure, CERN found the Higgs, but it's no God Particle! People "in the know" have so much as said that they found a black hole to hell instead.

KISS. Keeping it simple, it is easy to see that time has teeth, but when you tie time in a knot those teeth aren't as threatening are they? Nonetheless, time is the ick in the air. Time cuts. Space is a saver! The zero is the womb which carries the liquid solvent. *The soap*. The saver and the solvent would not exist without the cross. We can come out of chaos when we come into the cross. The cross unifies the four standard forces into One Standard Force. The cross, also, seamlessly explains the centricity required to maintain balance and to reveal the stabilization of gravity. The cross is central to life, literally and symbolically within its own construct of time itself. Is it a coincidence that Christ is placed in the center of our millennial timeline? Our calendars refer to Him in B.C. (before Christ) and A.D. (anno Domini (in the year of our Lord (Christ)). Academic attempts to change B.C. to BCE (before the common era) and A.D. to C.E. (common or current era) may be formal in texts but it is not catching on among the people. I've noticed most writers prefer to stick with the traditional reference to eras, myself included. The cross is central to life and the glyph itself represents the centric nature from which all formulas and particles must have been birthed.

I would propose that *Gravity* and *Electromagnetic* forces were/are found from the first quantum spark and thereafter *Weak* leaks in the balance discomposing it, giving us entropy. Disrupting the balance further are excessive (out of balance) electromagnetic waves as they exacerbate our atomic and cosmic atmospheres at rates ever increasing. I submit that the very first universal spark was the *Strong* big *boom* of the force as it met with the beam in unification. As soon as fusion occurred it created space. Space created a central body giving us gravity. A body has weight. Oh boy, here comes mass!

Chapter TWΣLVΣ

The Strong Arm

Although I claimed to have not had much love in my home growing up except when dad would tuck us in at night, I now see things differently, even with Oma. The scope of hindsight allows me to see her unselfish acts of love I could not see at the time because of the smite I had for her.

"How could a professing Christian make a fist and punch her kid?" I would complain. "Turn or burn" was her motto.

In all that God has revealed to me in my life, whether it be visions of Heavenly things or hellish things, one thing stands high above the rest. Love and forgiveness. This is the elixir for all soul wounds. Hate and resentment will fester into chronic fatal diseases and spoil the soul, whereas, love and forgiveness will heal a broken heart, a broken body and restore the soul to purity. It's not as easy as it sounds. I found it was much easier to hate than it was to love. Easier to resent than forgive. If anyone tells you it's easy, they're a liar! Or lying to themselves. If it were so easy,

God would have had no need of sending us a savior which cost Him everything. It was a verse in Luke 14 I took for granted and didn't examine how it applied to me. True forgiveness cannot be obtained until we first realize how much we need forgiveness. And we don't realize how much we need forgiveness until we realize how much it cost God to give it to us.

Forgiving Oma was one of the hardest things I ever had to do. The thing is is the bad memories return all the time and I am faced with the temptation to become embittered again. That is why the apostle Paul told us in 2 Corinthians 10:5 to bring every thought captive to the obedience of Christ.

Nowadays, each time a thought/memory of pain returns, I give it to Jesus and He is faithful to take it from me so I do not continue to think about it and become tempted to give into those memories and backslide into a heart of resentment. Forgiveness is not a one-time thing. It's a daily prayer.

Growing up, our pantry wasn't exactly overflowing with food. We struggled, as I recall my parents arguing about money. I remember on more than one occasion; Oma would refuse herself the meat on the table just so her children would have their fill. Then she would eat the leftovers if there was any. This is an act of love, not most people practice today. As for the beatings... I deserved punishment; I was not innocent. I may not agree with Oma's harsh choice of discipline, but I do take responsibility for my acts of disobedience. Oma's lack of affection, I know now, was more of a cultural thing. Koreans, I've noticed by and large, aren't very touchy-feely people, at least not in my family. A stark contrast of our Scandinavian grandmother who could not keep her hands off of us. Hugs and kisses all the time! I return to these memories to quench the bad ones. If I dwelled on the whippings, I would become angry and risk the love and forgiveness to be clouded by resentment, and therefore contaminate my heart. I want to move forward, not backward. I choose to remember the good times, like when my cousins, aunts and uncles (we have a large Korean family) would come over for holidays. Oma's brothers and sisters would lay down an oriental mat on the living room floor and sit in a circle to play Hwatu (aka Hafanuda, Japanese Flower Cards) for hours on end. This was after having eaten from a delicious Smörgåsbord of Korean food, of course!

I remember the love of my mother; I just have to get passed the dark clouded memories of pain. I remember the love of my father if I look beyond my own childish resentment. I have to move past the bad memories that want to dominate my mind. Once I find the love, I set anchor there! There was love in my life! More than I deserved! I know now, but even then... I knew love.

The consistent love I found in Jesus, and subsequently the steadfast love I found in my grandmother as well as the Pastor's family, showed me what this love ought to look like in human form. The Pastor's wife was another grandmother to me, and as a show of respect in Korean tradition, I addressed her as such. I called her "halmeoni" which translates to grandmother in Korean. From now on I will refer to her as Harmony (the English variation). She was a darling and dainty little lady with a smile and sparkle in her eye every time I saw her. She had a real warmth about her that radiated from her being. Anytime anyone was hurt you would take them to Harmony, and she and Pastor would pray for them. News traveled fast that people were being miraculously healed from cancer and all sorts of things. It was even reported that Harmony's handprint was left on the stomach of a dying woman in Colorado Springs who was miraculously healed from stage four cancer. Her handprint was proof that the fire of God was working through her. Harmony only had one hand, one arm for that matter. She lost her left arm from collateral damage in the Korean War. The one arm and hand she did have was used by God mightily, and more so than all the hands and arms of the whole congregation!

It was winter 1979 and all our belongings were loaded in the Ryder 18-wheel-moving truck. The Chevy Impala station wagon was packed too. My dad, and my brother Solo were in the truck ahead of us. Zeke (the baby, four years old at the time) and I sat in the back seat of the Impala wagon with Oma at the wheel following dad. We were moving to Denver from San Diego. The road trip was uneventful until we reached the mountain pass leading up to The Eisenhour Tunnel from Dillon/Silverthorne, eastbound on I-70. Today, I-70 has three lanes going up the mountain eastward from Dillon. I think it has, at least, that many lanes going down too. The tunnel, today, has multiple lanes also. In fact, it has separate tunnels for eastbound and westbound traffic. Not back then, east and westbound traffic shared a road and there was only one tunnel for both lanes headed in opposite directions.

We were only about an hour and a half away from Denver, making our way through the snow-covered Rocky Mountains when Zeke and I, holding each other in sheer panic unable to move, began to cry wildly at what we beheld.

"Come on! Jump out! Bali! Bali" Oma was shouting. Bali (pphal-li more accurately) in Korean means hurry. Zeke and I were frozen. We could hear her command, but we could not do anything but clutch onto each other more tightly. We were still in the back seat of the Impala, but Oma was not driving. The car was still moving, but now it was moving backward. We were coasting down the mountain pass while Oma was running along-side the car, struggling to keep up and hold on to her purse over her right shoulder, she opened the car door behind the driver's side where I was sitting. We were watching in sheer terror as Oma, crying wildly too, was screaming at us to jump out of the car as it was rolling backward down the pass, and the angel of death waiting for us at the bottom of the cliff.

Dad was not 100% on board with the decision to move to Denver. He was doing well as a salesman for Prudential Insurance Company in San Diego, after retiring from the Navy. We had a nice house in Imperial Beach and the weather in San Diego can't be beat, he would say afterward, regretting the move. He was strong-armed by Oma who would not take no for an answer.

Four years earlier Oma had a miraculous conversion from a life of smoking-drinking-gambling to giving that all up-cold turkey-and following Harmony wherever she went. Oma will tell you her conversion story began with a miracle which could not be explained except for the mighty hand of God at work. She saw a vision of both Heaven and Hell which shook her to her core and caused her to fall on her knees and repent of all her sins. And right after that a large mole above her left eye fell off like a dried-up peppercorn after a prayer conference where Oma said she and Harmony specifically prayed for the mole. She gave her life to Christ and became Harmony's right-hand woman. Or I should say she stepped in and became Harmony's missing left arm. Whatever you want to call it, her protégé. Her student. Her shadow. Oma was her loving assistant. She did whatever Harmony asked of her and followed her wherever she would go. Like the nursery rhyme, "...wherever Bo Peep went, the lamb was sure to go..." In truth, Harmony had Oma attend all her home visits as an appointed witness to the meeting. This custom was to deter false testimony and gossip. A practice more churches and institutions would be wise to follow and install as administrative protocol. Just say'n.

Oma insisted we move to Denver because the Pastor's family moved there to start a church, *and we must go help.* Reluctant, dad gave in in the end. *I think she was having panic attacks without having Harmony nearby. Harmony was soft spoken, unlike my mom, but according to Oma, Harmony emboldened my mom to be proud of her stern approach to life... that there was a place for her severe stance and a Heavenly calling for her firm fist.

"Bali! Bali! Jump!" Oma cried with both arms outstretched and her huge purse bouncing off her back as she ran to keep up with the car. I will never forget the look on her face. Pure panic, sheer terror, and fury. How was a seven-year-old and a four-year-old supposed to respond to that? Disobeying Oma always ended up in a spanking, but not this time. Once the Impala came to a buffering stop she grabbed us and held us tighter than Zeke and I were holding each other. That's the only memory I have of being in Oma's hugging arms. No wait, there was one other time.

Thank God for all the snow! The snowbank beside the road guarded us from falling over the cliff on the other side. If there had been a metal guard rail you couldn't see it for the snow. Either way, we were guarded. By Guardian Angels, no less, Oma would assert.

It was snowing and the road and entire mountain pass was blanketed by it. There was so much snow falling we could barely see the cars going by on the highway. It was a white-out. Barely able to notice as it was camouflaged, a man and wife in a white Econoline-type van pulled over and pulled us out of the snowbank and towed us to the top off the pass. Whether the Impala was driving too (while being towed) Oma can't recall, but our heroes managed to start the Impala for us, nonetheless. Just before the entrance to the tunnel is where we stopped and unhitched the tow rope. Oma hugged the couple goodbye with tears of thanks and we drove onward. The van remained behind us for safe measure. Oma told them that our car died and started sliding in the snow and she couldn't control it. She asked them to stay close behind us until she could find dad, who was nowhere to be seen. No cell phones in 1979. Too bad we didn't have a CB radio. That would've been perfect.

The Eisenhour Tunnel is 1.697 miles long, and as I mentioned, in 1979 there was only one tunnel shared by traffic going both ways. Halfway through the tunnel we saw the Ryder truck, dad laying on the horn, as he sped past us, now westbound on I-70 as we were headed east. I remember seeing five-year-old Solo standing up on the passenger seat hunched over dad to get a better view as we passed each other, waving and bobbing his head. Dad and Solo both apparently thrilled to have spotted us. Dad managed to find a place to U-turn the 18-wheeler at the Loveland Pass exit on the east side of the tunnel after he noticed we were not following behind. Now he had to do the same thing on the other side!

The following school year, my brothers and I were enrolled at a private Baptist academy. A small private school in a big Baptist church! We were students there for a year before dad ran out of extra money and had to pull us out and enroll us in public school in Littleton, where we lived. About that time, I became seriously ill and spent time in the hospital. The doctors didn't know what was wrong with me and Oma was too impatient to wait for a diagnosis. She insisted for the hospital to discharge me so she could take me to the healing hand of Harmony. A vessel for the healing hand of God.

I was in and out of consciousness, so, I don't remember much, but I remember enough:

I recall crying in pain at home in my bedroom. "It feels like someone's banging my head with a hammer," I told my parents, as I held my skull with both hands, sobbing. They kept me home from school and I was in my bed for many days with the same symptoms growing worse. I had a fever and I slept a lot. *In 9th grade I had Chicken Pox, and it was a lot like that in the beginning. In bed and in and out of consciousness probably due to the fever and my body's instinctive way to cope with the pain and trauma.

Same deal at the hospital; only awake for brief moments, enough to remember I was there. I do remember leaving the hospital. In my parent's arms. Whether it was my dad or my mom holding me I don't recall. What I do recall was being in Oma's arms as she laid me down on the front pew of the sanctuary at church, swathed in a blanket. *This is the other memory I have of being in Oma's hugging arms.

It was official. Pastor and Harmony from San Diego now had a bonified church and a growing congregation in Denver. Of course, being Harmony's apprentice, Oma had a key to the building. They rented the chapel of a Baptist Church off Federal Blvd. Next door was Winchell's, and I remember Oma trying to stuff glazed doughnuts in my mouth.

"Eat."

I tried. Then I threw up. Oma had to clean it up before returning to her fervent prayers, kneeling beside me. She was waiting for Pastor and Harmony to return from wherever they were. They, along with their two teenage sons, lived in a bungalow behind the church building.

It felt like we were there a long time. Alone. In the dark. Only a light from the lobby area illuminated the sanctuary. I fell back asleep.

I awoke again when I was being carried from the church to the bungalow. I was laid down, still wrapped in a blanket, atop a mat on the floor in the middle of Pastor and Harmony's living room. All three knelt beside me. Harmony asked my mom to un-swath me so she could lay her hand on my belly. The last thing I remember were those three faces; Pastor, Harmony and Oma squinting their squinty eyes as they prayed over me.

The next time I awoke it was bright. I mean, yeah it was daytime, but it seemed so bright, like the sun was shining directly in the living room. I squinted my eyes to shield it from the glare. I sat up from my makeshift bed, the swaddling blanket I was wrapped in, the mat beneath it, and an oriental rice grain-filled mini-pillow. I was still in the middle of the living room. I looked around as everything came into focus, and Harmony, with her bouncy and benevolent demeanor came over to me from the kitchen.

She looked at me, as if to examine me, smiled, and asked, "How do you feel?" She spoke perfect English. And Korean. And Chinese. And probably Japanese too. She was brilliant!

"I'm fine," I said, realizing after I said it, I really was fine. I mean, I was well. I was healed. I was back to normal!

"Praise the Lord!" She said as she raised her one arm in the air to give God the glory, "are you hungry?"

"Yes," I said realizing I was famished.

"You want some "lamyun?"""

"Yes, please."

My first miracle. First of many.

"Lah-myun" or "na-myun" is Korean for ramen. We didn't eat Top Ramen, mind you. Oh no, we ate the good stuff. Sapporo Ichiban! Lol!

I realize I painted Oma as a villain, but I did not do it to seek revenge, shame her or intentionally depict her as a hateful person. I stayed true to my testimony, that's all. To this day she does not realize her "spankings" were extremely over-the-top. I believe she believes she did the right thing to teach me lessons. Unfortunately, growing into a young adult it did more harm than good to our relationship. She was impossible to please. Still is, I'm afraid. Oma had always condemned me for being a hippie. My bohemian attire of oversized clothes and stinky soles from my Birkenstock sandals displeased her every time she beheld me, and she did not refrain from making her displeasure known. When I was 12, she traded in her wooden spoon for twig switches, and as an adult she traded in her sprig weapon for lecture lashing. Today, still she is unable to tame her tongue from stomping all over me. It is tiresome, I admit, but God sustains me as His grace renews me and melts away harbored bitterness for Oma. For anyone and everyone.

Harsh discipline was instilled in Oma as a child as her oma did the same to her at an early age she told me... a generational "tradition." Discipline is a continuum, in its own right, sort of. It moves both ways on a scale from zero to extreme and with proper balance it's a useful tool to keep us in check and to be accountable for bad behavior. *If you want ninja like qualities, whippings are a must. Except in my case, lol! Kinda wish I were a ninja.*

In the spectrum of discipline Oma took extreme measures to *teach* her children lessons, especially me being her oldest, when we were disobedient. As much as I oppose taking this extreme approach of discipline, I don't know what's worse; the other extreme of doing nothing results in some despicable attitudes and behaviors. It promotes children to become disobedient, self-righteous and entitled. It's obvious both extremes of discipline can be severely detrimental to a child. Lucky for me, I had Jesus to make up for damages and all loss has been reinstated in full and then some. I am truly blessed.

You know I no longer hold a grudge against Oma. My own healing and restoration would not have been realized had I not completely forgiven her but maintaining a heart of sympathy toward her is another story. It is not easy, yet, in the struggle God wins me over when He shows me Oma is battling her own demons. You know what they say, *we all need to walk a mile in another person's shoes before we pass judgement.* It softens my rigid stance, but then I tend to hardline myself again once I find that I am unable to yield to Oma's unending complaints and afflictions of discomfort and pain. Her distress calls are causing me anguish as she is long-winded in her descriptions of misery. *"Ooooh, my back. Oooh my eyes. Ooh my tummy..."* Her adversity is not reserved only by her physical pain, but her fluent grievance overflows into all manners of misfortune. To make matters worse, she is garrulous about it all. Coming together to pray is the best thing we can do for each other, but we don't do it enough, I'm afraid. Living with Oma is a roller coaster ride. The highs and lows can be nauseating. One minute I'm annoyed, then God will give me spiritual vision to ease my exasperation. Next minute I'd see her with compassionate eyes (an impossible feat without God's heart).

She was unloved so many times it caused her to misconstrue the love she had to give; so, when she was overcome with anger she took it out on me, usually. Jesus went through a lot worse, so I don't feel sorry for myself anymore. My only struggle is staying unbothered and maintaining my temper. Oma has proved to be unselfish in ways that matter; she has emptied her wallet on numerous occasions to help others in need and has refused herself to be the first in line when others are hungry. She will starve for the sake of others. Don't get me wrong, she still has some "whack" in her! She's been known to smack you upside the head from time to time, even in her old age. Lol! The Bible says our heart is attached to our treasure. Oma's income is less than $400.00 per month, but she always insists on picking up the bill when we go out to lunch. I'm told she does the same with others. The center of Oma's heart is a lollipop for others. She manages to give her tithe, and still support local and foreign missions, as well as buy groceries for the hungry. I am not attempting to justify her *spanks of discipline*, but from this perspective Oma is a saint... I guess *who could use some restraint*. Lol!

Discipline at times requires painful correction. The Bible says not to spare the rod. A balanced beam would make this rod effective as it would be anointed with love. The problem is love has been spoiled by human interjection. We cannot love properly without the love giver ruling our hand of discipline. I once condemned Oma for spanking me with sticks, but I've learned from the Bible that sticks and stones are less painful than words. Our tongues are powerful weapons which will steer the course of our lives more than any manmade weapon. The Bible is riddled with verses about the power of what we speak. When I realized God was revealing the importance of what we say and the words we choose, I began to take better care of my vocabulary. It's tough, I admit. Blurting out the F word is a hard habit to break. God will hold us accountable for everything we've said or didn't say and not just what we've done or didn't do. I'm not talking about cuss words. I'm talking about blessings, curses, and words rooted in love or hatred. It is imperative we rest in God all day during our day so we are not in danger of saying something we may regret. I learned this lesson the hard way, and still find myself falling short. I pray all the time that I have love rule my day. *Please God, don't let me be a noisy gong,* I would plea.

"If I speak in the tongues of men or of angels, but do not have love, I am only a resounding gong or a clanging cymbal. If I have the gift of prophecy and can fathom all mysteries and all knowledge, and if I have a faith that can move mountains, but do not have love, I am nothing. If I give all I possess to the poor and give over my body to hardship that I may boast, but do not have love, I gain nothing." 1 Corinthians 13:1-3

Oma's astringent carriage of herself could use some milk and honey to sweeten her conveyance and to raise her pH level, (I'll admit, for me too). In spite of her untamed tongue, she has proven to me that she would readily lay down her life to protect mine. It took a long time but now I see the center of Oma's lollipop. She's a lover, not a hater. She just picked up some habitual behaviors for which she continually suffers from.

It's my duty as her daughter to go through her mail, help her pay bills, decipher documents, and reconcile her accounts because she depends on me. Her 2nd husband is unable or unwilling to help her and my two younger brothers are not available. These responsibilities do not deter me from loving her. I don't even mind her criticisms of how I dress or look. The battle I face with Oma, spiritually, is the temptation to resent her when she comes to me with continuous complaints. Furthermore, in regard to her health, ironically, her faith is lacking when it comes to believing in Jesus for well-being.

Years ago, I was diagnosed with rhomboid and infraspinatus muscle spasms. After college I would regularly get episodes of this condition which could last for several weeks. I would be hunched over, unable to stand straight until the spasm would loosen its grip. I took muscle relaxers regularly which didn't help whatsoever, and I underwent many therapies including traction. Oma's oma, my "harmony," had severe osteoporosis and lived the last two decades of her life as a hunchback. I feared the same fate for me. Until the day I realized the power of the spoken word. In faith, I would silence my painful moanings and complaints by speaking healing over myself in Jesus's name. It worked almost immediately. I have been totally free of muscle spasms for over ten years. I shared this with Oma to encourage her to do the same with her own scoliosis and numerous other body ailments, to no avail. She continued to voice her complaints and to this day it seems as though it is all she talks about.

The consistent grievance she unloads on me is like noisome pestilence. My attempts to effectively pray with her have been stonewalled by her unbelief and persistent need for a physician to prescribe her more medication. She protests the supernatural power of God in her own body while boasting of His miracles in others! One day, I pray, she will receive the body-transforming miracle she desperately desires and, ironically, *knows how to receive*. She must first transform her mind and free herself from a generational curse which is preventing her from the physical healing she desires (for which I have already begun to pray about). We are to bring every thought captive to the obedience of Christ before we speak it into the universe. Our reality is governed by our mouths. It pains me to write all this. All of us, believers or not, need awareness that God is able and willing to give our lives transformation. If we only had faith. Why are so many of us dead set in our self-destructive ways? Deep down we know what's wrong and we do it anyway. I'm speaking for all of us, including me.

My ability to love Oma today, unconditionally, is nothing short of a miracle, more so than the physical healing of my back or the myriad of other miracles I've had. I've experienced many miracles. As a witness to others and personally. These experiences adhered to me like nothing else and has cemented my faith in an all-powerful healing God. From my first miracle to my *last*, each one acts like another layer of foundation. Not much can shake my tree of faith anymore. It's grounded. It's rooted. I still seek medical help when necessary, but it is rare. God has been so good to me. *Oh, when I say my *last* miracle, I do not mean the final one, I just mean the last one to date.

If I had to rate my own personal miracles it would be difficult because they were all amazing in their own right. But, I definitely have a top-three list. *One*, you know how it ended; with a bowl of lamyun. *To be clear, when I say *one*, I do not necessarily mean that one is number one on my list. I just mean it is *one of* my top three.

Another one (miracle), you do not know how it ended, but *you do know* how it began. On a *snowmobile*... in a ditch... stranded beyond Irwin lodge...

I'm sitting on the stump, crying my eyes out while looking down at my vertical snowmobile in a ditch. Left knee throbbing. The good knee. I'm taking swings at the air. A boxing match with my invisible doppelganger. Myself. *My bad self*.

"You are so f**king stupid, Gwyn!" Followed by two jabs and a left hook. "What the h*ll is the matter with you! You f**king idiot!" F bombs everywhere. While I was cussing, screaming, and punching... something caught my attention. It was Him. He came as a tender voice. Like a warm breeze and light as a feather. Gentle. I couldn't see Him, but I heard Him.

"Look up." He said.

What?" I froze. I held my breath.

"Look up," He calmly said again. It's hard to put into words, but the voice was almost tangible, like a soft current or gentle wave. It enveloped me.

I looked up and before my eyes toward the west, the clouds split open and the late day rays of the sun poured out like a waterfall. If you recall the day was overcast with thick stratus-type clouds. It was an incredible sight to behold. I was stunned. I was silent. I was not alone.

"Get up," came His voice again. It was more of a command. Not like a drill sergeant or anything, but like a tone that held with it confidence and tremendous courage. So much so, it instilled those feelings into me. Out of nowhere, I was full of confidence and courage!

God didn't say anything else, nor did He have to tell me to get up a second time. The determination and fortitude that overcame me led me. He did not give me any further instructions, but I instinctively knew what to do. Without hesitating I got up and marched down to the snowmobile, not once feeling the pain of my injury. As soon as I reached the edge of the creek, I grabbed hold of the snowmobile with both my hands and pulled that thing out of the trench in one fell swoop.

Incredibly, there was no damage to the front end. No dents. The front skis weren't bent. The windshield was not cracked. The keys were still in the ignition, so I gave it a turn and it started up immediately. I put on my helmet, straddled the beast, and made my way back to the lake. As soon as I got there, it was time for departure. To the minute. Kris waved me over as she waved the others over, as well. When we finally grouped together, I wanted to tell them what happened because it was so fresh, and I was still pumping with the adrenaline of the whole ordeal. It was a miracle! I just experienced a GINORMOUS miracle! As excited as I was I tried to shout over the noise of the engines.

"You guys won't believe what happened to me!" But no one could hear me with their helmets on and motors running. Besides, no one was paying attention to me. And no one knew I was missing in the first place. I flexed my biceps, reliving what just happened. I can't even begin to explain the unspeakable emotions flooding my soul at that moment. I wanted to cry. Maybe I did. I wanted to sing. Maybe I did. I wanted to shout from the mountain top! I wanted to jump off the machine and dance around the lake on my HEALED good-bad leg! But I didn't get the chance because we were moving out.

Recalling the event of that day, so long ago, it's just like the other memories where everything after the fact is faded. I don't remember anything after that. I don't recollect what we did at the cabin, or the ride back to the cabin. I don't recall anything. I must have been on Cloud *Nine* for the remainder of our stay, because I sure don't call to mind anything else. God won the day!

"Illuminous beings are we; not this crude matter."
-Yoda, The Empire Strikes Back

The Process of Illumination

Everything and everyone are included.
You and me,
Trees and bees,
Stars and snow,
Ticks and fleas

Atomic systems,
Systemic actions,
Cosmic rays
And
Quantum fractions

The waxing moon,
The whining baby, The deepest hole,
And darkness... maybe...

Light shines in the darkness, yet darkness has no light...
which brings us full circle back to the paradox in sight,
The paradox of Might.
This Might be that and that Might be this
But what Might really matter Is the Might of KISS

2007

Chapter TH1RTΣΣN

Light of the World
Part I

Where there's fire there's smoke! Following the metaphor I used regarding the cross, sign of smoke means something's cooking. All matter must come from the magma of the spark. The energy emitted from the point of fusion would make weight of something because the birthplace of space would give rise to everything. Every cross would bring forth a new generation of weight. In 1997 an article was released announcing Stanford University in collaboration with Princeton, University (U) of Rochester, and U of Tennessee created matter out of light. The article reminded readers how converting matter into energy is common, but the reversal of that has been a struggle to accomplish. Their experiment with their SLAC atom smasher revealed "virtual photons" creating electron and positron particles. https://www.slac.stanford.edu/vault/pubvault/tip19902000-/tip1997-dec.pdf

Their experiment proves my point. The spark of the cross is the original spark of all life! Is it so hard to imagine the formulation of a 3D or multi-D world from the cross? The cross is a cartesian coordinate in of itself, after all. Geometrically, a tesseract gives us a cube. Or consider the zero which gives birth to a cone and if inverted diagonally, it reveals a square! Just say'n. All of life in all its shapes come from the cross! All of light in all its hues, come from the cross!

$E=mc^2$ means we have energy and its mass is moving at the speed of light. I've suggested that mass was created at the same time as light. + = fire= light= heat= space= energy= mass. If all these things can be unified into the cross how does it account for the velocity of light? Over a hundred years ago the Michelson-Morley Experiment revealed the absence of an aether. An aether in the atmosphere was to provide an explanation of how light moved in waves. Like a ripple in a pond, the aether, hypothetically, was the pond for waves of light. Although their experiment failed to prove the aether existed, another experiment by Georges Sagnac seemed to prove otherwise. There seems to be a rising argument over the validity of the Michelson-Morley experiment. Regardless, I propose light waves are not dependent on any type of spatial fabric within our three-dimensional space because the place and space of the fabric originated from the light, right?

Dr. Einstein combined the formula of kinetic energy to his shrinking factor and arrived at the most famous equation of all time. There *should be* a pun here. He recognized their mutual affinity for each other when experimenting with the idea of moving at the speed of light. I encourage you to read Dr. Michael Guillen's book, Five Equations that Changed the World to learn more about Einstein's shrinking factor. He explained the shrinking factor and its merge with kinetic energy beautifully.

If space is the *noun of life* and time is the *verb of life*, it only stands to reason that anything that moves in life is pushed by time. Light. It moves.

Nothing could move within space as we know it without it because space is made of time's cross section. Time is sectioned in space as we know it; measured by our seasons and tracked by our clocks. Inside light we can see time unadulterated. Uncut. "Un-sectioned." The knot has been loosed. The *pinhole* of the cross allows its force to come through the gate. *All light (visible and non-visible) include radio waves... the entire electromagnetic spectrum.

What I've come to realize about light, particularly the speed of light, the constant speed of light, is that its velocity of 186,282 miles per second is not how fast it goes from point A to point B, although it has been measured to move that fast (with the exception of moving through crystal fluids). Instead the velocity pertains to the speed of its movement within the light itself. A ———→ B. The time it got from A to B is not what I am referring to. If you were to stick a speedometer in the beam itself, after it had already reached point B you would find that the light was still moving at the speed light. What does that tell us? It tells us that light is in constant perpetual motion moving at 186,282 m/s even after it hits its target. It is indeed a constant and a continuum of energy. Light contains the space time continuum. Its internal waves prove something is moving it along. Time is pushing it. When fire/light was originally ignited by time[2] it began to move at the speed of time at all time. The speed of light is the speed of time. I'm not talking about how we track time with our sexagesimal clocks. In light of my vision I'm resolved to say that time itself moves that fast... minus, maybe, the drag variable. Light/heat may have some weight and entropy to it.

Speaking of entropy, where does *entropy* come in? The exhaust of the spark? Time is the ick in the air. The opposite of the usable good stuff, right? And then what about the *poop* and *pee of the particles*? The soot of smoke? It only makes sense that the waste product of these particles would have to be detected somehow. The excrement could be discharged as ick in the air! Maybe as negatively or neutrally charged particles? I know this and I'm willing to bet on it: Time is entropic. *Weak* leaks of the force, aforementioned, have teeth and are deadly. Strong force has even bigger teeth. Electromagnetic force is an eater. And gravity is lethal as well. It all comes from the same beast with four different heads. They all eat. They all have teeth. Time is a knife. It cuts. It eats. We've established that, right? The four forces are unified into one. Wouldn't it make sense to unify the force even further into time itself? It couldn't be more obvious, right? The only thing that stops time, or can pinch it shut, is zero in reverse. The circle closing into a single point. The gate closes into the cross. It is the Kevlar armor to defend us against the bite of the beast! Einstein unified our existence to time and space without realizing it. $E=mc^2$. E = existence, m= space, c^2= time. God is the One Standard Force, therefore, He is in control of our existence as He is the true light bearer, holding the key to time and space. The cross.

The cross unifies everything, except for maybe one thing. Two things. *Three things*. Poop, pee, and entropy! Or the poop and pee are measured as entropy. LMAO! Genuinely, I have faith the cross will unearth, unwind, and blow the minds of scientists as they start to apply it in their own research. Everything the cross reveals is intentional. Everything it builds is calculated from zero to infinity. Its design is fundamentally demonstrative as the first stitch in the fabric of life. The first stake hammered in the ground. The first brick. The cornerstone and the capstone. It holds it all together. It is the foundation. The unmoving rock of ages.

If you ask me or any other creationist, all life is deliberate, not an accident by random particles, raising the question of order out of chaos (unless we're made of soap, lol)! Are we made by lumps of carbon or nucleic acid which got lucky meeting each other as it floated in nothingness? The mathematical odds of that are so huge evolution theories cannot even put a big enough number on it as to how long ago that would have been! On the other hand, our double helix circling a common axis known as DNA is proof an architect was at work. There is a complex blueprint in every human being and every living thing. The construction of these processes is not nailing pieces of random planks together. The finished product speaks for itself. "We are beautifully and wonderfully made!" I echo the Psalmist with the same reverence and awe he had for a creator, but my heart is heavy in knowing that mankind is in a pursuit to renovate it all.

Genetic engineering is fueling mankind's ego in thinking that we can reshape our bodies without consequence. It begins with good intentions. Remove mutant DNA. Set a course for controlled evolution, so we become stronger and more adept human beings. Do we really think that playing cards with our chromosomes will result in a jackpot for mankind? No no... the house always wins.

In my three-day vision, I saw atoms unable to crash into each other. They respected each other's space. Even the ones that were joined in their positive and negative magnetic pull, they had mutual regard for each other. They never violated each other's "personal space" because it was the natural order of their relationship. Well, they did "hug" each other, but they never went to third base if you catch my drift. Two atoms do not make a baby. At least, not in my vision. A new generation of atoms is made within the womb of the light particle. Or at least the light particle carried the "seed" or the egg. Maybe that's what photons are doing when they flicker in flight between the nucleus and the electrons. Maybe they're pollenating! I dunno, but I do know that when you force atoms to fuse to make a third, it is unnatural, and the repercussions are destructive.

When I began to read about atoms and their immense power it quickened me. The thought of man attempting to harness the energy from them did not excite me. It gave me the worst sense of foreboding which led me to my knees. Why would mankind consider to even try to crash atoms together? What in tarnation's is wrong with us! I appreciate how Werner Heisenberg gave us his Uncertainty Principle to understand

Quantum Mechanics, but, I'm disappointed by how he also had a hand in developing the atomic bomb, which made me mad and sad.

World War II came to a close in August 1945 with the atomic bombings of Hiroshima and Nagasaki. Inasmuch as anyone, as a student I was happy to know a horrific war came to an end, but the development and past use of this type of weapon does not restore hope for humankind.

Both atom bombs used in Japan were fission bombs, but they were each different in their delivery method and at their core. The bomb used on Hiroshima was called Little Boy and the one used on Nagasaki three days later was called Fat Man.

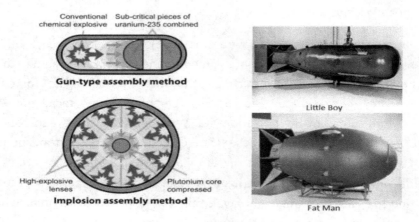

56. Atom bomb types, 57. Little Boy and Fat Man

Little Boy had uranium 235 (U-235) at its core. U-235 is an isotope. Isotopes are radioactive elements. You recall from chapter two; these are *slow* nuclear *spillers*. You have uranium 235, 238, 239, etc. Each one has a different level of instability and radioactivity. Little boy was ignited by a 2nd mass of U-235 as it fired into it like a gun. That is why Little Boy was considered a "gun bomb" and it was slender in shape. By comparison, Fat

Boy was just that, fat. It had plutonium-240, although the developers were wanting P-239 because P-240 had lower energy output. They got around this by constructing a different ignition system. Instead of firing one "bullet" at the core, they surrounded the core with a firing squad, which they called the implosion assembly method. This second bomb which hit Nagasaki was called Fat Boy not just because it was round in shape, but it also would do more damage delivering 21 kilotons of force. Little Boy exploded with 13 kilotons, but killed more people, estimated at 237,000. Fat Boy killed an estimated 80,000 total. Not because it defected; the lower number of casualties was because Nagasaki evacuated the city beforehand due to another raid which was happening in the area.

Most of the people evaporated leaving behind footprints where they once stood in the black soot of the ground. Others were not so lucky. They burned to death, trying to find relief in nearby rivers. And yet still many more died slow deaths from the radioactive fallout (the left over uncontained weak nuclear force). Those who survived tell stories of appalling details. They all agreed it was hell on earth.

*To learn more or hear testimonies from survivors visit the Atomic Heritage Foundation at
https://www.atomicheritage.org/history/bombings-hiroshima-and-nagasaki-1945

Fat Man and Little Boy were fission bombs. Fission, as opposed to fusion, splits in half. The technique to cut open the atom was to cut it in half with another atom of the same kind, thereby causing a chain reaction for more atoms to split open. All releasing its strong nuclear force. The copilot that delivered Fat Man was quoted saying that when it ignited the "light of a thousand suns illuminated the cockpit" and he had to shut his eyes even with dark welder's goggles on.

Since then, scientists pushed the atomic envelope and have created an even heavier bomb than Fat Man. The Hydrogen Bomb is a fusion bomb, where they join two hydrogen isotope atoms (lithium deuteride) together forcing them to birth a third atom. This atomic rape releases its fury. Get this, in order to ignite this bomb, they must generate enough heat to fuse the two isotopes. They use plutonium or uranium as a spark plug and drive it into a "supercritical" state to start a chain reaction by fission. Essentially, they put the old atom bomb inside the new H bomb in order to ignite it!

Over 100 million kelvins of heat and pressure during the fission stage and up to 300 million kelvins of heat and pressure is generated in the fusion stage to ignite the bomb.

*Fun Fact: 100 million kelvins = 179,999,540.33° F x 3 = 539,998,620.99° F. That's more than half a billion degrees Fahrenheit!

Hellfire. Not really a *fun* fact, is it?

This beast of a bomb will deliver 100,000 kilotons of destruction and vaporize everything in a five-mile radius of ground zero. All the matter; every last cell within the body will incinerate.

I can only imagine what it does to our atoms. When does the nuclear chain reaction cease?

See https://en.wikipedia.org/wiki/Thermonuclear_weapon

Pierre-Eugène-Marcellin Berthelot, a 19th century French chemist was quoted saying, "Within a hundred years of physical and chemical science, men will know what the atom is. It is my belief when science reaches this stage, God will come down to earth with His big ring of keys and will say to humanity, 'Gentlemen, it is closing time.'" Ironically, Marcellin was an atheist, but even he knew, then, the "God-huge" implications and consequences of playing with atoms (quote taken from nytimes.com).

Tampering with atoms is playing with fire the likes of which we cannot even fathom. Again, I ask why would mankind think it's ok to tinker with atoms? For "free" energy? Is the energy contained within them worth it? It doesn't sound free to me. Sounds like there is an extremely high price to pay. Are the particle accelerators, aka atom smashers (like The Hadron Collider at CERN) made of the same material as these bombs? I mean, these bombs are mini atom smashers, are they not? How do you contain energy and heat levels at that magnitude? How do these scientists sleep at night?

Remember, Stephen Hawking (before he passed) and the pope have both said that CERN is knocking on hell's door. CERN's high-tech underground laboratory is like a ticking time-bomb. And if it doesn't explode, something else will come through. I once saw a testimony on YouTube of a guy who claimed to have had a vision from God. He said that Jesus or an angel of the Lord took him to CERN, and he saw a black mass come out of one of its chambers and cover the whole land.

Whatever comes out of that place will not be good. Just like a portal (ie, the Ouija board, stargate), it is opening a door to the other side. A big door! If you ask me, all nuclear power plants and accelerator labs should close their doors, especially CERN. What's the point? The power and/or technology harnessed... is it worth it? Why is it still operational? Is it because CERN gave us the world wide web? *And the world wide web is glorious. Don't want to shut it down! We cannot live without the internet!* By the way, W in English is the Vav in Hebrew and the Vav is the number 6. Is it another coincidence that www is interpreted as 666 in Hebrew? The mark of the beast is everywhere. It is so predominant these days that people scarce take it seriously. The doomsday clock is at three minutes to midnight people. Wake up!

"You must accept Jesus, or you will go to hell," was the first thing Oma would say to friends I would bring home. I stopped having friends over. Oma, still to this day, is convinced that I wouldn't bring friends over because I was embarrassed of her Korean heritage. Our Korean heritage. That just was not true. Her broken English, or the smell of kimchee or doenjang (concentrated miso) permeating the air in our home were not the reasons I was embarrassed. I was humiliated by her unwavering command to commit my friends to hell if they didn't drop to their knees, right then and there, and repent of their sins. *Who was she to judge?* I resented her. Oma did all she could to rear us in the Christian faith, but she was not helping the situation. As aforementioned, I am a Christian today, not because of the tender-touching-loving role model she was... not... traditionally... but in spite of it.

Funny how things change. Just like Oma, God have me a vision of hell although different (hers was a place, mine was hellfire), but it had the same effect. Oma says all the time how we must follow Jesus to save us from hell. I no longer find those words offensive, and I chuckle to myself when I say something similar, recognizing the irony of it all.

I have been called a conspiracy theorist, but more accurately, I am a conspiracy factualist. I take note of the facts behind every theory and if the facts are lacking evidence, I lack interest. Yet, if facts are lacking it doesn't mean I write it off entirely, especially if the coincidences are too obvious to turn a blind eye to because many times a mere coincidence is the first clue. Plus, some facts are spun from fiction. Also, God has given me visions and spiritual discernment regarding Satan's plans, and He quickens my spirit when I hear truth. Satan's schemes are stranger than fiction and it is often worse than the theory! Many conspiracy theories out there are partly true and from what I've uncovered these secret schemes are more sinister and perverse than rumored and worse than I could have ever imagined. But it doesn't surprise me. The Bible warned us. The pieces are lining up. I do not take coincidences for granted, and it is mind-blowing to recognize the times we're living in with real-life events syncing with

Bible prophecies. The world today. It is Biblical. And not paying attention to the things that are most obvious will only leave me blindsided.

We are on a runaway train with humankind's employment in the atomic realm. Moreover, man's effort to manipulate genetic codes and rewrite them stretch to the outer limits of our natural world. Mankind just needs to get his hands out of the micro/nano/atomic cookie jars. There is nothing delicious in there. If it looks appetizing it is a ruse. It's poison! What about the science of medicine? Without our microscopic discoveries we wouldn't be able to combat disease. That's a good thing. While that may be true, why is all of medicine based on a symbol that represents the serpent? Satan. The caduceus? Call me superstitious, but this gives me pause, especially when I see it in the lap of a Baphomet. I get the creeps. Is there a better way?

58. Caduceus of Hermes, son of Zeus 59. WHO logo, 60., 61. Caduceus in the lap of the Baphomet

Anything with a snake in its symbolism causes me to 2nd guess its purity and what the symbol stands for… really. Don't get me wrong, I have nothing against snakes, per say, its roll in its natural habitat (and I actually like them and do not have some phobic aversion to them), but when it is used to depict the roll of a manmade construct it doesn't settle well with my soul. It tells me there is some underlying malicious motive at work. Why? Because the Bible tells me so. I'm not against doctors nor nurses. I am not making a generalization condemning medicine, nor all who work in the medical field. God has used medicine and medical personnel to help people, I do not deny. I've just come to realize that Satan has his hand in every industry on earth and he loves to show off by putting his stamp on every one of them. Once you start to connect the dots, you will too. He is so proud! No shame. Satan's symbolic snake is one of many, and in this specific case it is linked to the tantra of the Kundalini spirit of "divine feminine" energy.

God has shown me so many things through the years which I kept to myself mostly because of the outrageous nature of it all. But the whole purpose of this book is for me to be completely transparent about everything so others may know the truth and break free from Satan's deceptions regardless if some people naysay what I say. A couple years ago God showed me there was a snake coiled around Oma's spine. Terrible, I know! When I learned that Hindu tradition welcomed the kundalini spirit in Yoga meditation to do exactly that (entwine the spine (like the number *nine*)) I was appalled. Then when God showed me Oma's spine (spiritually), He revealed to me that non-Hindu people have the same snake coiled around their spine. They essentially invited the same serpent spirit unknowingly when they became dependent on medicine. Satan aims to control us so what better way than to get his claws into our nervous system? The sneaky snake that he is, he uses good things such as health and exercise to slither his way into our lives. If he can drive our thought processes then he has a sure chance of ensnaring us and keeping us blinded. Chemical dependence is contractual with malice.

The Kundalini concept of Hinduism involves a snake wrapped around the spine beginning at the bottom vertebrae/chakra. It coils up the vertebral column in succession with the person's height of enlightenment as their third eye "awakens" and the snake reaches the top of the spine with its fangs (I imagine) plunged into the pineal gland. The pineal gland is responsible for opening the third eye. Does this remind you of something? The Eye of Horus? The apex of the pyramid on our dollar bill? Why does Osiris (father of Horus) pose with a staff with a pinecone (symbolic for pineal gland) flanked on top? *And why, pray tell, does the Pope's staff copy the same symbol? Just say'n. The pineal gland is known for regulating our internal rhythm (circadian clock) as it produces melatonin. It also regulates hormonal gonadotropins from the pituitary gland which controls sexual development. Hmmm, I wonder what the snake of "divine feminine energy" has to do with our reproductive organs and sex drive? This snake, God showed me, is one in the same as in the emblem stamped to modern medicine. This is much more than an abstract symbol!

Is it a coincidence that pharmacy gets its name from the root word *pharmakeia*, Greek for sorcery? Doesn't the Bible warn us of sorcerers? Ancient sorcerers dabbled in chemicals, made potions, and most likely opened the first drug stores, apothecaries. Today's sorcerers of Satan may not be at our corner Walgreens or CVS, but we can bet our last pill that pharmaceutical laboratories are crawling with them... right next to their lab rats! Not just over-the-counter or prescription drugs are we served from their menu of medicines, but their *remedies,* I'm sure, reaches internal hospital care of individual intravenous injections to weapons of mass infection. Their anecdotal recipes for antidotes, antiseptics or vaccines, tinctures and tonics morphed their way to ammunitions. Serums originally made to help us have transmuted in the hands of sickos into toxic concentrations of chemicals. Agent Orange and Mustard Gas, and God knows how many other nerve agents were created in their laboratories. Gaseous, liquids, and viral concoctions are, no doubt, continually added to their menu of items. Less threatening homeopathic remedies are trending, and surely, Satan, in order to get a piece of the holistic pie, is poisoning this "safe" alternative any which way he can. Beware, "all-natural" does not guaranty no contamination.

Drugs (some, not all) are more keys for the doors of our spiritual gates to open which allow snakes to come through. Ick! Heard of Flakka? A synthetic drug causing people to act crazy at a new level of insanity. Some have dubbed it the zombie drug as people have allegedly tried to bite other people. Reports of super-human speed and strength are common. It takes numerous police officers to apprehend one person. Testimonies from people who have taken the drug claim it felt like they were possessed by the worst evil imaginable. I recall some bad experiences tripping on shrooms or acid, seeing Satan manifest from a Robert Plant poster of Led Zeppelin when I was a senior in high school, but nothing the likes of Flakka. Thank God; it was horrible enough. Are hallucinations not simply the third eye opened to an unseen realm? Pagan teachings, worldwide, would not argue with this statement, and apparently modern medicine with its serpentine ties wouldn't either. According to Jesus, mental illness, regardless of its cause, is simply demonic oppression or possession and can be cured by invoking His Name. Wow, could it really be that easy? If the core of modern medicine is aware of such supernatural occurrences why do they relabel mental illnesses with a sleuth of modern terms warrantable for persons exhibiting those illnesses to a facility lockdown? Looney bins! And then pump them with [more] drugs! Does this sound sane to you? I realize there are positive results for many taking psychotropic drugs, and it may truly help some people, but I cannot help to wonder what Satan's sneaky, sleazy, snaky hands are doing!

"Gwyn, you're over-reacting," my husband would tell me. But I'm not. I'm not reacting at all. In fact, I play right along just like everyone else. Although I have these suspicions about medicine, I still go to the doctor, albeit rare. I still get check-ups. I take prescribed allergy medicine and still take anti-biotics and other meds when they are prescribed to me. I am in the game. I have not withdrawn from it. I have not "over-reacted." If I choose to refrain from conventional medicine and conventional technology I would have to join an Amish community. So, in fact, I am not over-reacting one bit. It just amazes me that we so-called Christians depend on medicine and turn to Jesus only when medicine and treatments fail. The cool part is is that Jesus, full of compassion, meets everyone at their personal level of faith and we still see miracles occur. The modernization of high-tech medical procedures and medicine does not deter people from praying to Jesus and asking for His supernatural intervention (even if it is after-the-fact), nor does it deter Jesus from answering their prayers. This is just another example of His amazing grace.

We look upon their lifestyle with disdain. We, the modern world. The Amish people, and even the lesser strict Mennonite's, are considered old fashioned or passé. They live like Puritans. *Who wants to be pure! It is ridiculous.* Yet, more and more people are choosing to live off grid and return to simple living. Not to be pure, no no, but to be ready for the coming apocalypse! Even they can sense something on the horizon. But most of them do not have to rely on their sixth sense because the writing is on the wall. It is plain for all to see. If you do not see it you are not paying attention.

The idea of a bright future for our shining modern age with its flashing billboards not being all it's cracked up to be is catching on and is becoming quite trendy these days. I see more and more survival ads and militia-type paraphernalia for sale. All of us are catching wind of the notion that we are all on a thin line and everything as we know it will fall off. EMP. World-wide pandemic. Nuclear holocaust. Alien invasion. Asteroids. Natural disasters caused by climate change, and economic collapse. The list goes on. There is no shortage of Hollywood films dramatizing each scenario. Yet, it is curious to me that like the number 666 found everywhere we look, the idea of an apocalypse has become mundane to the general population. For the majority, we just go about our business, planning our holidays, and going to the movies like all is well in the world. We ignore all the warning signs, but deep down we know our society is skating on the edge of an abyss. We choose to look in the direction of the rising sun and pay no attention to the setting "artificial" sun behind us and the smoking pit it sets under. The dark ages may be behind us, but the darkest hour is yet to come.

We tend to overcomplicate things. Explain things away. As adult humans we are especially good at it. Jesus said you must become like a child to enter into the Kingdom of Heaven (Matthew 18:3). What did He mean by this? It seems pretty obvious. Children are innocent. They do not have preconceived ideas. They view the world with awe and wonder. They're pure, pure and simple. They are the true puritans of our world, yet, ick is in them too. Even the Amish, free from tech, is not free from ick.

After I mop the floor, I let others know not to walk on it. Of course, I do not want them to slip and fall from the wetness, but the main reason is because it is clean, and I do not want them to dirty it up. Once it's dry I ask them to check the bottoms of their shoes or take them off. I want to keep the floor clean for as long as possible. In fact, I have been known to not drive my car for a week after I have it cleaned for fear the weather outside will ruin the shine.

In the same fashion God will not allow anyone to enter His home without first taking their shoes off. He has higher standards than I do. He has the highest! We must be pure. Free from any trace of ick. Our shoes must be removed before we cross into those pearly gates, so the shoes can be thrown into the fire with all the other ick. Otherwise we're liable to contaminate the whole place! Children understand this simple explanation of God's goodness. Yet, we feed them new age garbage of maintaining a positive outlook on life when, in their own strength, they're just as weak as we are! They need to draw their "positive" source from the only positive source. That goes for all of us. There is no amount of positive thinking [meditation] I can achieve to clean my car. The only positive vibration my car needs to remove the dirt stuck to it is from high-powered hoses and/or brushes. I have to go to the source. Because we are three-part beings, body-soul-spirit, we must, likewise, go to the source for three-part cleansing. Is repentance the remedy for ick-cleansing? It all begins with what we eat.

In chapter three, I quoted Hebrews 4:12. "The word of God is living and active…" remember? I have experienced firsthand why it is called the *living* word, and it has changed the course of my life. It is written: He (Jesus in unification with God the Father) was/is the word.

> In the beginning was the Word, and the Word was with God, and the Word was God. He was with God in the beginning. Through him all things were made; without him nothing was made that has been made. In him was life, and that life was the light of all mankind. The light shines in the darkness, and the darkness has not overcome it. John 1:1-5

The Bible is the Bread of Life, that's why we are supposed to read it daily… to feed our soul and spirit. But so is Jesus… the bread of life. He became the living word while the Bible remains the written word. He said in Matthew 4:4, "It is written, "'Man shall not live by bread alone, but by every word that comes from the mouth of God.'" And He said in John 6:35, "I am the bread of life; whoever comes to me shall not hunger, and whoever believes in me shall never thirst." And in John 6:51 He said, "I am the living bread that came down from heaven. If anyone eats of this bread, he will live forever. And the bread that I will give for the life of the world is my flesh."

And we all know this one from the Last Supper recorded in all the gospels and further made famous world-wide by Leonardo DaVinci:

> While they were eating, Jesus took *some* bread, and after a blessing, He broke *it* and gave *it* to the disciples, and said, "Take, eat; this is My body." And when He had taken a cup and given thanks, He gave *it* to them, saying, "Drink from it, all of you; for this is My blood of the covenant, which is poured out for many for forgiveness of sins. "But I say to you, I will not drink of this fruit of the vine from

now on until that day when I drink it new with you in My
Father's kingdom. Matthew 26:26-29

Jesus is the living word (the *logos*) and the word is the bread of life. If
we eat of it we will have life eternal. Is it as simple as that? Medicine for
our soul? Does it clean it and make it pure? Also, if we are unable to gain
access to a Bible to feed our soul and spirit, will Jesus keep us from
spiritual starvation? Is it possible that if we had perfect faith in Jesus,
modern medicine would be unnecessary, as His body (the bread of life)
would keep us perfectly healthy? Without ick we would never get sick? Or,
if we are sick from ick, all we have to do is say sorry? Repent? That easy?

What is pure? What is not? Why is purity controversial? Either
something is pure or it's not. It is not a dubious concept and it is not
disputable. It is plain. Purity itself is the proof and the dirt, the ick, is the
evidence. The proof is in the pudding, and in the end we will all see what is
stuck to us. The ick will speak for itself. The ick is what Jesus called sin.

Ick = sin

Sin is taboo. It's one of the most hated words in the western world.
Why? Due to its negative connotation, it has become politically incorrect
to mention it. *We hate that word. And we hate Jesus.*

Why is Jesus so hated by the world? Isn't it curious that Jesus alone
has been targeted by mainstream as some bad guy? Any other "god" or
idol is ok, but as soon as you mention Jesus all hell breaks loose. What is
up with that? Why are Christians singled out? Because they love Jesus,
that's why! We either love Jesus or we hate Him. Then there are those
who believe "Christ-consciousness" is a benevolent way to embrace Jesus
as one of many spiritual hors d'oeuvres on a tray full of metaphysical
options to ingest, like Buddha or Vishnu. *Let's sample them all*, they say! In

this case, Jesus isn't hated, *but in this case*, it's not the *real* Jesus. Many sincerely believe Jesus was just *one of the many,* and that He was, himself, a believer and a teacher of this credo. But this could not be further from the truth. Jesus said so himself. He said what He gives to drink will end all thirst and what He gives to eat will end all hunger (spiritually). He does not point to himself as one of many appetizers. He said he was the only thing to "eat." He so much as said He was the main dish. He said He was the one way to the Father and all others are poisonous. He broadcasted with His public following that He was exclusive. The priests of His day hated Him because He said He was *one with God* and the only way to God. I am hated for loving Jesus and spreading the good news of His Pure, One and Only Light from God, with God, as God. If you love Jesus as God you will be hated too. But why? I've heard stories from people who've admitted to hating Jesus and avoided and even hated Jesus lovers... they said they had a strange aversion to Jesus (as exclusive). They claim their revulsion was an unexplained antipathy toward Him. They couldn't quite put their finger on it. It was as if the opposition was otherworldly. I was curious to know why Jesus was always targeted as the bad guy. Never Allah. Never Buddha. Just Jesus.

Jesus said Himself, He is the ONLY way. The road is narrow. Jesus warned us, "If the world hates you, keep in mind that it hated me first." John 15:18

Why does the world hate Jesus?

Because He's the one with the ball. Everyone wants to sack the quarterback. The world at large hates Jesus because he is exclusive. The world wants options! There is no other way to Heaven, according to Jesus's own lips, "I am the way... no one comes to God but through me" and the world hates Him for it. Think about it.

He's not the bad guy. He's the *only* good one! If you read the stories for yourself you will see (for yourself) that He was all things lovely. All things true. All good things you want for yourself and for those you love. If anyone rejects this notion they have not read it (for themselves) and have been duped by haters. The story of Jesus is lovely. It doesn't get any lovelier. Seriously, what is there to hate? A baby born in a manger surrounded by adorable animals and a host of angels singing in the Heavens under the brightest star? Even sorcerers from Persia brought baby Jesus gifts. Even they knew a king was born! It is written in recorded history (in addition to the Bible) that Herod The Great ((Herodes Magnus, appointed Roman king and ruler of Judea) killed every first born child under two years of age in Bethlehem in his evil quest to kill the king prophesied to come. Little did he know that Jesus did not intend to become king of Judea, or king of this physical world (at least, not yet). All those poor babies died in vain. Jesus was hated before He was born. ☹

Jesus worked with his hands as a carpenter in his early years and worked with his hands as a healer in his *latter* years. He had the ability to command a revolt and incite violence on the Roman rulers as well as the pompous priests. He had so many followers, he had to "escape" or retreat in early hours of the morning to pray by himself. It was the only time He could be free from the crowd. Can you imagine? An exhaustive onslaught of people wanting your attention. He was the "it guy" back then. There was no celebrity more popular than he. And you know what, he became their consultant for free. He became their therapist for free. He became their preacher for free. He became their physician/healer for free. What a great guy! The best! He could have done all this to raise money, but He didn't.

He, also, could have raised an army; so many of his followers hoped so! In fact, many "disciples" left his side once they realized he would not tolerate violence because they only followed him hoping to overthrow the Romans. So, they left His side because Jesus was soft. Too loving and forgiving. Only twelve truly devoted stuck with Him. Even so, Jesus could have called on the legions of angels that were at the ready to do His bidding. It is written one angel alone conquered 185,000 Assyrian soldiers in 550 BC. And let's not forget what Jesus said to Pontius Pilate, the 5th Roman Governor of Judea under Emperor Tiberius... just before Jesus' crucifixion:

So, Pilate said to Him, "Do You refuse to speak to me? Do You not know that I have authority to release You and authority to crucify You?" Jesus answered, "You would have no power over me if it were not given to you from above." John 19:10-11

Jesus could have overthrown the oppressive Roman regime if He wanted to. He had the manpower and the "Heaven-power." But He chose to be the humble servant. Before he was betrayed by Judas, arrested (at the Mount of Olives) and handed over to the high priest, Caiaphas, and his cronies, He and his 12 disciples were in the "upper room" celebrating the Passover. The lovely story continues... Jesus washed his disciples' dirty feet. You gotta understand, they wore Birkenstocks back then! Lol! They wore sandals. Their roads were not paved. They literally had dirty feet. Jesus knew this was the last meal He would have with them as He told them so. Jesus wanted to leave one final impression on them for them to remember Him by:

> When he had finished washing their feet, he put on his clothes and returned to his place. "Do you understand what I have done for you?" he asked them. "You call me 'Teacher' and 'Lord,' and rightly so, for that is what I am. Now that I, your Lord and Teacher, have washed your feet, you also should wash one another's feet. I have set you an example that you should do as I have done for you. Very truly I tell you, no servant is greater than his master, nor is a messenger greater than the one who sent him. Now that you know these things, you will be blessed if you do them. John 13:1-17

This most lovely story does not end... After they left the upper room they went to the Mount of Olives, particularly to the Garden of Gethsemane to pray. This is where blood dropped like sweat from His brow as He prayed. When the Roman Guards arrived to arrest Jesus, Peter drew his sword and cut the ear off one of the soldiers. Jesus, in His love and compassion, healed the soldier and restored his ear whole. It must've been a sobering moment.

During His interrogation, Caiaphas and the rest of the pharisees spat in His face because He claimed to be the Son of God, and they were plotting to arrest Him for this reason. When questioned by the priests (at a time long before His arrest) about who he claimed to be, Jesus responded, "before Abraham was, I AM." The priests wanted to stone Him then, and I'm sure gagged and tore their clothes, so disgusted, as Jesus boldly positioned Himself on the same footing as YHVH! Equal to God. Jesus had rebuked them time and again because they were "holy" in appearance only. They loved being called Rabbi, teacher. They were proud. Jesus told them over and over, that they made a mockery of God. They refused to help the poor man on the road who was beat up by thieves, and it was the Samaritan who had mercy on the poor man and stopped to help him. The Samaritans who they (the Jews) hated... who the Jewish high priests hated and led by example.

I must digress. Just so you know, in the old testament, God brought judgement on nations, including Israel when they neglected to help the poor, the homeless, the widowed, the orphans. For those of you who believe that God in the Old Testament Bible was depicted as being cruel, you're wrong. Yes, his wrath is to be feared if *you* are cruel. Sodom and Gomorrah were destroyed by fire and brimstone. You may know the story. Sodom is where we get the word sodomite. Most people think that Sodom was destroyed because God hates gay people, but the truth is Sodom was guilty in all acts of sin, not only sexual immortality. The story does not point to homosexuality exclusively either (Jesus loves gay people as much as anyone. To say otherwise is a lie). But for the record, to God, sex is sacred. God hates sexual immortality (hetero or homo) because it desecrates the sanctity of a holy union. It's dirty and it's selfish. *Let's keep it real, most people will be eternally separated from God due to unforgiveness and lack of repentance for being haughty, never mind all other "sins." Truth is, most of us are unwilling to get on our knees because we are deluded by selfish desires. There was so much self-indulgence in Sodom and Gomorrah it caused them to neglect people in need, no doubt. God put an end to that. What's more... the [other] cities the Hebrews conquered and plundered... were they not occupied by *non-pure* humans? The Bible says they were giants. Does not Genesis six clue us in? More and more Christian scholars now concur they were likely the progeny of fallen angels who seduced and had sex with (many raped, I'm sure) human women as confirmed in the book of Enoch. Did Nephilim and/or Rephaim species result in those unholy unions giving rise to and passing the genes of giants/Anakim (as recorded in Numbers 13)? Were they the Titans of Greek mythology? The Anunnaki? Norse gods? Egyptian alien reptiles or Apkallu? Reptilians of Aryan, Draconian or Saurian origins? The gods of Babylon? All of the above, I'm sure. Did they "mix" genes with mankind and animal-kind to give birth to myths of ancient creatures like Centaurs and fairies, as well as legendary monsters (and contemporary "cryptoids" like the Wendigo, sighted and documented by modern man)? *Cryptoid, my cousin informed me, is the *word* for creatures of urban legends, fyi. To the ancient Greeks they were known as chimeras. Anyway, the book of Genesis says they were impure, and Noah was the only perfect one. Surely, the Bible is not referring to Noah as being spiritually or morally perfect (that would contradict the Bible), therefore, God must be referring to something else. The genetically impure? Ancestral forefathers of the cryptoidal kind? In light of the Genesis 6 account it is reasonable to believe everyone except Noah had genetic mutations of *unGodly intentions*. Noah was the only thoroughbred of God's "image and likeness" left in the world. God had to preserve His bloodline for the coming Messiah. Jesus could not be born from a gene pool contaminated with miscegenation of species. According to the book of Enoch, demons are disembodied spirits from these hybrid species. The seed of Satan found its way back in the world after the flood (postdiluvian era) and began "mixing with kinds" once again. Perhaps a dormant gene was carried by one of Noah's daughters-in-law or perhaps a disembodied spirit inhabited a corporeal body and the crossbreeding began once again. Some speculate the Tower of Babel was a portal which allowed men to mingle with interdimensional spirits after the flood. Others maintain that the Nephilim returned and descended on Mount Hermon. Regardless how, the hybrid giants, again, spread and took over the region (we know of today as Israel and Jordan (then worldwide)) before the Hebrew nation took it back after their famous exodus from Egypt. The Hebrews killed and plundered many towns of these giant species at the instruction and anointed power of God. Don't feel sorry for them. They were alien (not pure human) offspring and they were cannibals as well as maneaters. I'll bet Sodom and Gomorrah were infested by them. God said He would spare the entire city of Sodom if Abraham's nephew could find just ten righteous men among them. He could not find a single one. God was looking out for the good guy. Each time He destroyed cities and nations He did it out of His love for us. To spare us from the infected. They were like parasites, out to destroy all that He had created. A "cordon sanitare" wouldn't suffice. You cannot quarantine a nation-wide infection. Unfortunately, they arose again. Generations later, the Bible tells of how David slew the Philistine, Goliath, with one stone (of five) slung to his head. Time and time again, God wiped them out... for you and for me. So, all who have believed that God was cruel and unjust for destroying cities and nations with fire and the world with a flood are wrong. God is for us, not against us. He is cleaning the ick.

On with the story... Jesus was lovely in all ways always. After He was flogged and forced to carry a tree on His back they whipped him some more as He made His way down Via Dolorosa all the way to Calvary. But they didn't have to crack the whip as He went. He was not going to go astray. Like a lamb to the slaughter, He would be obedient unto death. At Calvary, the place of the skulls, He was crucified. For you and for me. The place of the skulls was called Golgotha, a word originating from Goliath (that horrible man-eating giant). I've heard atheists say if Jesus was God, then He wouldn't have felt any pain and it would have been no big deal for Him. Jesus could have healed himself on the cross and kept himself from bleeding out. Jesus could have healed himself from the whipping post, beforehand, by closing all his open wounds, but he didn't. He certainly had the power to avoid the pain, but he allowed it. He had to suffer, for our sake. If not, the entire sacrifice would have been for nothing. The pain came as a consequence, and if you ask me, the level of pain he must've suffered was probably worse because of His unhindered level of endurance. He was hard to kill, I'm sure. It may have taken more whippings than for the average man to get Him to weaken. I bet it took more time to kill him because of his strong bright light and purity. The light/life that was in him would've been hard to stifle/kill, no doubt... but it was done, and He said so himself as He drew in His last breath, "it is finished."

Even during his suffering Jesus found compassion for the guilty men hanging on the cross next to him and was able to save one of them as the thief submitted to Christ in his final moments while he still had breath in him. When Jesus's blood was finally poured out and a knife plunged into his side to be sure, it is written, darkness came over the land (an eclipse?) and there was an earthquake. It is written, the curtain (9cm thick... actually two curtains according to Jewish tradition, 9cm x 2) of the temple (that separated the Holy of Holies from the lesser) tore miraculously in half... symbolic to the entire prophetic event of the life of Jesus! Now the only thing standing in the way to the Holy of Holies is Jesus himself!

The story doesn't end there. The best is still to come. There was one Rabbi priest (who was not one of Caiaphas's cronies) who believed in and adored Jesus. His name was Joseph of Arimathea. He asked Pilate if he could have the honor to bury him in his family tomb, knowing Jesus was essentially a homeless man with no retirement/funeral fund. The Roman King agreed and also agreed to appoint Roman guards outside the tomb at the behest of the other priests. After Jesus's body was taken down from the symbolic place where He gave up His spirit (a place made to memorialize maneaters) he was ceremonially cleaned and wrapped, and His wasted carcass was placed in Rabbi Joseph's tomb.

On the third day, Mary Magdalene and "the other Mary's" went to the tomb to pay homage with their spices and was surprised to find the stone rolled away and the chamber empty. Two angels, who glimmered like lightning, appeared to them and asked why they were looking for the living among the dead. The angels reminded the ladies of Jesus's words to them while they were in Galilee just before that final journey (for Jesus) to Jerusalem, "'The Son of Man must be delivered over to the hands of sinners, be crucified and on the third day be raised again,' Then they remembered." Luke 24:7

Jesus appeared to them, and to the 11 disciples (Judas committed suicide), and to two on the road to Emmaus, as well as 500 witnesses it is written, before He ascended with a host of angels into Heaven, blessing them as He went.

Those left watching with their eyes open not only to His ascension but to the scriptures He had shared with them pertaining to all messianic prophecies and the written word being fulfilled as the living word among them, were themselves filled with the hope of His return as He had promised beforehand, "And if I go and prepare a place for you, I will come back and take you to be with me that you also may be where I am." John 14:3

This is the same promise made to the bride by the groom in Jewish tradition. And so, the church today is considered the bride of Christ waiting for His return to take "her" to their new home. Come Lord Jesus!

Here is God's amazing grace: While God would have been justified to destroy His creation, He decided to blaze a path of redemption for us, instead. Think about it. We cheer on the hero in the movies when he/she takes revenge on the one(s) who harmed his/her child(ren). We sympathize with their revolt, and even when they murder the perpetrator. Contrast that with our attitude toward God. Do we cheer Him on? Our attitudes toward Him wound Him. His heart breaks. We put salt on God's wound when we mock His crucified one and only begotten son. When we spit on Him with sarcasm and hatred. We scourge Him all over again with our unbelief and crucify Him with our hate. He would have been justified in taking vengeance on us. Rather, He extended more grace. We do not deserve His mercy, but He gives it to us anyway. Correction: He gives it to those who KNOW they need it, and approach Him with a humble heart, sorrowful for sins, and repentant. The poor in spirit. The poor in spirit are opposite of avengers.

You would think that in light of His story (aka The Gospel) everyone would agree that Jesus is worthy to be praised. That it would render Him as *the only one* worthy of worship by comparison to all other men or gods in all of history. Who can compare? Anyone? Anyone? Who is purer? Who is more deserving? He willingly shed His divinity for you and for me. He surrendered his kingship in order to fix what we messed up. Who would do that! He became a lamb. Died as a lamb. Resurrected as the King He was and is, but also now as the high priest qualified to redeem us. The risen Lord will return, but not as a lamb. When He comes back He will return as The Lion of Judah, as foretold by the prophets of old. Why aren't more people holding signs on street corners that read, "The end is near. Repent. Jesus is coming!" But even so, people would pass them by, not at all affected, like the boy who cried wolf, everyone ignores him... but the wolf does come in the end. Not so in this case, the ever-so-deserving King Jesus, crowned with glory and honor, comes to (once and for all) close the rabid jaws of the wolf and bring light, justice, peace, and purity to a dark and dirty, spun out world full of self-loving nyctophiles who commit unspeakable sins in the dark and "speakable" sins in the day.

Growing up, Oma would always remind everyone that Jesus was coming. I had always cringed when she would say that because I could not accept it as a possibility, not in my lifetime, at least. I now share her sentiment. I've realized whether He does or does not come in my lifetime, my spirit unified with His, joins in the hope and song of this exclamation! *Come Lord Jesus* is the longing hope of all believers. *Maranatha!* To once and for all have our corruptible flesh and our imperfect hearts and minds changed to an immortal spirit made perfect by His perfection as we are *joined to* His incorruptible Body and His everlasting Kingdom.

Sickeningly, it turns out that the most loveable human who ever lived is the most hated man among mainstream culture. In fact, His name, Jesus, has become an expletive. A curse word. Does this make any sense? We see a man carrying a sign warning about the Day of The Lord, and we wince, and turn our eyes away from the outcast, and use Jesus's name in vain. *We hate him. We hate Jesus.*

Is it because His light shines so brightly it makes us all look bad? We can't stand the spotlight revealing on us the inch thick of ick! Could it be because He says come as you are, but we are unwilling to take off our mask? Maybe it's because He holds the bar *too low? I don't want to wash my neighbor's feet! Gross! I don't want to serve others! I deserve better!*

The angels the day He was born, sang, "Glory to God in the highest..." We would rather have them sing, "Glory to man in the highest!" We say, "my will be done," instead of, "Thy will be done."

Why? Why is this the human condition of so many? For as many as there are people who hate Jesus there may be just as many who claim to love Him but are not doing anything to hold back the wave of hate. They think they are, but instead, they are joining the pool party and making a mockery of the immaculateness of the Messiah. The mainstream subculture of "churchianity" is not helping matters when it comes to reflecting the image of Jesus. Unfortunately, it broadens the gap of hate toward a stainless Jesus and makes a fool of His followers. Their view of love and unity is contaminated with world views of what is and is not acceptable. They really believe they are acting in "love" as they mingle in the mud. Instead of shining Jesus's light they are filtering their own. Jesus called us to come out of the world and counter the culture... to blaze a narrow trail among the wide path of popularity. If Jesus were here today, too many self-proclaimed Christians would crucify Him all over again, I'm afraid. He would disrupt their comfortable "Jesus-as-a-crutch lifestyle! To boot, most of the Jewish priests and scribes didn't recognize Him (as the prophesied Messiah) 2000 years ago and neither would most of the priests and scribes today! In doctrine, they didn't and still don't hate the Messiah. They know who He is in the *written word*, but they fail to recognize Him as the *living word*. Priests are still pompous today. What has changed? If Jesus were to show up as a servant today they would pass Him by as the scum of the earth. I'm afraid most of the churches today are not His bride. They are harlots. Many unclean spirits, along with Kundalini, have entered the sanctuaries of the church buildings and have also entered the sanctuaries of the human heart and have beguiled the bride of Christ. They have been escorted to another suitor without even knowing it (maybe some do). Either way, they are cheating on the groom. They are fornicating with someone else. They're having a secret love affair.

> He was in the world, and the world was made through Him, and the world did not know Him. He came to His own, and His own did not receive Him. But as many as received Him, to them He gave the right to become children of God, to those who believe in His name: who were born, not of blood, nor of the will of the flesh, nor of the will of man, but of God. John 1:10-13

LOVE-AGE CONTINUUM

Timeless love and love in my space
is as the northern wind of yesterday
kissing the southern wind of tomorrow,
creating a storm
with the recipe of attraction.
A passion from the creation of a body like Venus
dancing around its love for the sun.
Undivided by age as it only plays
as a physical framework of celestial spirits
from what beginning and to what end?

I am in love with one
who if were here
would be two thousand years older than I,
but whose essence is young
and eternal in time.

1999

Chapter TH1RTΣΣN

Light of the World
Part II

The main reason Jesus is hated by the world at large is because of something. Or I should say, rather, because of somebody. Satan.

Angels and demons in the unseen realm are subjects most people avoid, even by many of my "Christian" peers. I get it. Most people do not want to talk about things for which they do not understand. It's scary... the demon-devil part of it. Why do churches skim the surface of the topic, turn the page, and move on? I find it disappointingly ironic that I have had more profound conversations about Satan with non-church-goers. With all the paranormal television shows hitting primetime, "devil discussions" are becoming more widely accepted, unfortunately, to the secular audience more so than with Bible believing friends. Oma warns me to not give the devil power by talking about him. I see where she's coming from, but she is missing the big picture. Twenty years ago, this may have had some truth to it in regard to the attention I gave the devil, but today not whatsoever.

This is business. Serious spiritual business. Evading Bible verses which are deemed controversial seems to be the status quo among too many conservative evangelicals, I've found. How does the saying go? "Keep your friends close, and your enemies closer?" Is this cliché an overstatement or is it wise to be privy to the secrets of those who aim to hurt us? Nowadays with so many evangelicals compromising Christ-centered teachings with popular culture, politics, and pagan practices, can they now be called *demongelicals* instead? Everywhere I turn, the truth is twisted and true deliverance ministries from "demonic" influences are hard to find.

From an early age I sensed that there was a world among us no one could see or detect, like a circadian clock within me ticking to time and space unseen. This other world... it may be *classified* as unknown or undetected but, for me, I did know, and I did detect a force... there's a good one and an evil one. I remember when I was in kindergarten walking home from school, I was being followed by something unseen. I never questioned the validity of it, although I always wondered who it could be. My guardian angel? Maybe. I never saw it, but I sensed it was there... an otherworldly force who knew how to manipulate the four forces in our seen realm... I instinctively knew it, when at the time I had no idea what the four standard forces were. From that time on, for me the unseen forces did not diminish into myth, but instead, were confirmed in me by life experience. Then I began to really study a book which further confirmed what I already knew. The Good Book!

Spirits are real. With the good comes the bad. It's not all angels and silver linings. Like the vertex parabola, there is a positive and a negative side. This is also true of a force. An opposite to everything. Matter and antimatter. Evil is real. Satan is real. The Bible says so. I began to realize these forces could manipulate more than energy. Intelligent spirits, both good and bad, were forces diametrically opposed and capable of human influence. I had discernment early on but lacked self-esteem and self-control, so I was an easy target for the bad forces to take advantage of me.

The irony of it all is that while I lacked self-esteem I was confident and conceited in my ways, and while I lacked self-control I convinced myself I could not be controlled by unseen spirits. This is a dangerous mindset to be in, especially, if we are not protected by God. Thank God, I was graciously guarded (undeservingly) from falling into pits to hell, but I was foolish and still fell in many potholes.

Satan hates Jesus according to scripture and Satan hates mankind. The Bible tells us that Satan was the most beautiful angel in Heaven before he turned evil. It also tells us that he was the angel of music. The musician of all musicians! He must've been the choir director. The conductor of the orchestra. The band leader. The head guitarist. The lead violinist. The composer of opera. You name it. He has infiltrated the music industry; you can bet on it! My friends and I waited in long lines for hours (all day and overnight sometimes) to get the best possible seats to all the shows for which, I'm sure, the devil orchestrated.

Satan is more than the god of music and media. The bible warns he is a snake, a serpent, a dragon, a beast, a devouring lion just waiting for someone to devour. We also read that Satan is the Father of Lies. And he is a tempter. He is our adversary. He is the Prince of Darkness and the prince of demons. And Jesus said Himself, that Satan is the enemy, the evil one, a murderer, and the prince of this world. The apostle Paul echoes these sentiments of Jesus, and in 2 Corinthians 4:4 he says that Satan is the God of this World.

The Bible says Satan masquerades as an angel of light. It's no wonder, then, why numerous "Christians" are two-timing Jesus. Satan comes off as shiny and irresistible. A secret love affair? While I was cheating on Jesus, I did keep it secret the best I could, but for many, they flaunt it, perhaps not realizing it. Satan is the light of the world and the only light bearer according to his worshipers. The Bible says Satan's light is fake. It's counterfeit. He probably uses mirrors to make it look like the light originates from him. The Bible strenuously warns he will deceive the nations. Think about that. Let it sink in. Is it any wonder there is a global push for a blended religion? Tolerance? Sounds good. Sounds humane. But it dilutes the truth about Jesus being exclusive. If we compromise and forfeit the sovereignty of Christ alone we are subtly deceived by evil.

Evil is not a concept made up by religious zealots as Satan would have us believe. Evil is not an abstract idea, or an ambiguous hypothesis. It is not conjured up in the twisted minds of the mentally psychotic. It IS the twist in the mind which conjured up the evil to conceive the psychotic. The monster is not *the evil*. It is *the evil* that makes the monster. Is it merely a coincidence that the palindrome of LIVE is EVIL? The opposite of living is dying. Death eternal is dying to evil. To all of you devil worshipers: If you think you will head into the afterlife, happily ever after, with Satan as your king you are in for a rude awakening when you find that his kingdom is without light or anything living and you are stuck in a place much like Hiroshima and Nagasaki were when the place of the atoms around them have no containment in space! When the light is extinguished you will find yourself caught in the flaming exhaust of time, without any hope of escaping it. Do not be deceived. Nothing can *LIVE* there.

The Bible warns us that Satan will seduce us to do his bidding. Being tempted alone is not evil... it does not cause the ick to stick. It's when we give into it (even in our thoughts). By dwelling on *the temptation* do the devil and his demons have a right to torment our thought life, and by acting upon the dirty deed do they then have a right to stick their claws into us. And when they do they will not let go. Come hell or highwater, they will cling even tighter. It is their mission to drag as many souls to hell as they can. The only thing that can loosen their clutch is the hand of Jesus. The word of Jesus. The word is Jesus. There is power in His name. Even Satan himself will flinch and recoil at the sound of His Name. Don't take my word for it. Try it for yourself the next time you feel as though "something wicked this way comes" just call on Jesus. He is a savior now in this life as well as the afterlife. Listen to the testimonies from people the world over who have said so... Google it. The name of Jesus is like kryptonite to all unclean things and the blood of Jesus is like bleach!

I have a dear friend who could (and still can) see in the spiritual world since she was a child. She could see angels. But she could see demons too. When she was older and more mature she realized that the monsters would flee simply by calling on the name of Jesus. She was forever grateful and became a believer after that. I have heard numerous testimonies of people (with and without spiritual sight) who have experienced the same type of rescue from beasts of burden who have battered their lives physically and emotionally. Ironically, healers practicing Reiki needed healing from demonic attacks, mental anguish, and/or disease and physiological dysfunctions from fungus and microbes. Monsters show up in all forms and sizes. From the smallest to the biggest. Parasitic monsters are no less destructive than giants. Microbial and viral pestilence is growing. Even further, there is a rise in new plagues of formication. Fibromyalgia and Morgellons Disease didn't even exist in medical journals until recently. Tactile "hallucinations" are more real and prevalent now than they ever were. Praise God, so many have been delivered by Jesus, not just the feeble and poor in health. Mystics, Satanists, atheists, you name it, Jesus saved them all! There are numerous testimonies of people who once engaged in transcendental meditation and as a result became haunted by entities and many even claimed to have seen what they describe as demonic looking creatures. Many more claim these familiar spirits had morphed from beautiful beings or even relatives or loved ones who at first appeared to be clothed with light... and they had trusted them as their spiritual guardians. They testify Jesus alone saved them.

Even alien abductees by the hundreds have been able to ward off their abductors by calling on the name of Jesus. The aliens flee! Demons flee! The parasites flee! Testimonies abound! There is proven power in His name! Furthermore, a lot of regular people are having dreams about Jesus and dreams of the rapture and the tribulation. Just like me, according to their testimonies, they can tell a regular dream versus a dream from God... there is a distinction. Globally, people are being visited by Jesus. Google it. Whole Muslim villages in the Mideast are coming to Jesus for the same reason. Aboriginals in Australia, by the thousands, are coming to Jesus. Most moving of all, perhaps, is the underground church in China, where thousands are worshiping for 24 hours straight, tears streaming down their faces, overjoyed by the love of Jesus!

Our pastor made a pilgrimage to rural India recently. He reported that the people walked for hours to come hear the word of God. They packed the building and then begged him afterward to not stop preaching. Our church, who has mission fields in Peru, record countless stories in our newsletter of children and families who walk for hours to congregate for church. They all love to praise the Lord, Jesus! The world over, nations are unifying as they fall in love with Jesus and long for the quickening wind of the Holy Spirit. The one and only Ruach HaKodesh (Hebrew for divine force or Holy Spirit). They are realizing that when we allow God to use us as vessels He fills us with His water of life to purify us, feeds us with the Living Word to make us strong, and anoints us with Heavenly Oil as we worship Him. Like shampoo and conditioner, He cleanses us, but the oil, He showed me, can overflow our cups, and burn long and strong with His Holy fire to protect us from the flames of wickedness and hell on earth.

Unfortunately, simultaneously, especially in the western world, many people are seeking, via magik, witchcraft, enchantments, divinations, sorceries, games and cards, psychic hotlines, etc., to invite unknown spirits into their lives. They pay for psychic readings and counseling, séances, or some even go so far as to spend much money to enter sensory-deprivation chambers in order to meet death just so they can experience transcendental awareness of a parallel dimension or a future unseen or make contact with a loved one who has passed. All mediums to contact the dead are a form of necromancy which is strictly forbidden by God. God does not command us to refrain from these things to be a killjoy. He makes these commands because He knows that these things will kill our joy. Many "normal" things today which we are deceived to believe will bring us joy and peace, such as contemplative prayer and meditation, are invitations to the spirit world as doors to our consciousness are swung open with a welcome rug. The Holy Spirit will never enter in this way. All other spirits will. Ick.

We all seem to chase after supernatural experiences and fall for magik tricks as the solution to our dilemmas. Those who have admittedly been misled to believe the spirits in their lives were benign, which were later revealed as monsters, were tricked. The seemingly benevolent spirits were wearing costumes of elegance, charm, glamor, and allure. They were seduced by their exquisite ensemble. Satan is not ignorant to the things that enchant us, nor is he baiting us with one lure. His dressing room is as big as earth itself and his wardrobe is drop dead gorgeous. Hehe. This is no laughing matter, I know. The Bible says he makes war against us and with the heavenly host. Would he not send in spies disguised as "friendlies" to infiltrate unsuspecting humans who are willing to sell their souls for stardom? Today, people are lining up to sell their souls for a lot less. And many more are just giving their souls away for free. They hate Jesus that much!

More and more people are testifying to the fact that they used some form of magik to attempt to overcome or break the "spell" of a curse. *They used magik to fight magik.* This has been made popular and normal by Hollywood films. In the movies, the "good" magik wins. In real life, this is preposterous, and people are coming forward with their real-life testimonies of how Jesus was the only permanent solution to their haunt, ailment, curse, whatever. No manmade ritual could do it. We can't fight evil with "less-evil" and expect to have victory.

While many have come to the foot of the cross because they had a *Jesus experience,* many more are coming to Him because they had a devil experience. They were haunted or tormented in dreams or in real life and realized evil was for real. They reasoned, if pure evil exists in entities and as an intelligent force then pure good must also. They found through their own experience the only thing that will overcome evil is non-evil. From their testimonies they tried manmade options and "good" rituals, but they found true deliverance from Jesus. Period. Jesus cleaned house! Imitations may lather up and get sudsy, but the real solution to clean up the mess is not hidden under bubbles. The holy oil from our Heavenly father is the only solvent we will ever need.

Furthermore, I find it curious and ironic how westerners in droves are taking up Eastern religions as they are "enlightened" to it while the easterners who generationally practiced these enlightening religions for centuries and millennia, in droves are abandoning their Eastern faith and traditions and turning to Jesus because, as they will testify, He shines brighter. All the while, many modern-day charismatic Christians are diluting God's power with Eastern rituals, and most are unaware. Their erratic behaviors of being "drunk" in the spirit, I'm sad to say, may be nothing more than parlor tricks by devils. If we are not "discernful," our hunger for signs and wonders will be fed by intercepting spirits on assignment by Satan. Everyone is so hungry for the supernatural they are not diligent in testing the spirits as the Bible instructs us to do so. We want it so badly that we compromise the miracles of God, ironically.

We place too much credence and weight on subpar experiences. With stars in our eyes we sometimes convince ourselves we had a "God experience" when, instead, we were tricked. We unwittingly "pit" ourselves in danger. We inadvertently opened a door to the dark side. I've personally experienced (and I've seen in others) the breath-taken moment when a foretold event (ie, psychic who tells us the unknown) comes to pass. We are starstruck and hooked as the *sooth*sayer did its job by *soothing* our souls. We turn to horoscopes and tellers of fortune and then wonder why we suffer mental anguish. God's children have no business in dabbling with mediums or worse, acting *as mediums*. Now more than ever, Christians need discernment because the lie is twisted in so much truth, we need God's GPS to navigate us through the deception at every corner of blinking billboards built to distract us from our destination.

Satan has a pond stocked with hungry fish and it is spilling over with those chomping at the bit. At this point his lure has turned into a labyrinth of plucking LED's. He scoops them up as phish in his *net* by the billions. His life-lackluster web is worldwide and the lacework sparkles with neon lights, glitter, and gold. God's gold. If we want to commune with a pure spirit, there is one way. One formula.

Satan hates absolutes. He loves the gray area. Satan has gone so far as to dilute languages. He has inserted Dog Latin into our English literature and documents. See https://youtu.be/bfKNf65DjZM He's even going so far as to dilute math. If he can convince us that 2 + 2 does not equal 4 then he has got us, hook, line, and sinker! He wants to convince us that everything is relative. Sadly, it's working. It seems almost everyone has thrown common sense out the window and has come to a consensus that there are multiple truths. They love the gray area. Jesus calls us to come out from the shadows into the light.

There's more to the *Lovely Story*... in fact, it is the *neverending* Story.

"Jesus wept." The shortest verse in the Bible is found in John 11:35. Lazarus had died and was already rotting in the tomb for four days. Mary and Martha, and the other Jews, were disappointed Jesus didn't get there sooner to heal Lazarus and keep him from dying. Jesus wasn't late. He was right on time. He knew what He was doing. No doubt, Jesus was moved by the tears of those he loved mourning Lazarus as He loved Lazarus too. But

Jesus wept because of their lack of faith. He wept for the same reason he wept when He rode into Jerusalem on a donkey for the Passover Feast in the renowned Triumphal Entry. He knew, *this Passover, He* would be the sacrificial lamb. As He approached Jerusalem He wept, it is written. He said, "If you, even you, had only known on this day what would bring you peace— but now it is hidden from your eyes. The days will come upon you when your enemies will build an embankment against you and encircle you and hem you in on every side. They will dash you to the ground, you and the children within your walls. They will not leave one stone on another, because you did not recognize the time of God's coming to you."

Jesus weeps over us when we are trapped by the snare of Satan and do not recognize the treasure freely offered to us and readily available. He is able and willing to provide us with the key to unlock the chains that bind us. The bulldozer to break down the walls. The snips to cut the snare and the sword to cut the net of entanglement. He is able to free us and then willing to offer us RESTitution for the rest of eternity. Jesus says we are unsuspecting, and He tells us how He wished He could gather us to Himself like a hen gathers her chicks under her wings, but we refuse. Like the rebels that we are, we run into trouble head-first instead of heeding our parents' call.

The resurrected Jesus said, "I am the Living One; I was dead, and now look, I am alive for ever and ever! And I hold the keys of death and Hades." Revelation 1:18

This lovely Jesus, full of love and compassion, wants so much for us to choose Him. So he can purify us and make us clean in order for us to go with Him to be with the Father. To save us from Hades. He wants so much for us to hide under His wings, because only He can save us from hellfire. But He will not force us. He is a gentleman. The choice is ours and ours alone. Like Jesus, we should not force our opinion or religion down other people's throats. Our tactic to convince others of Jesus's love and power to save should be exactly that. Love others and reveal His power, thereof, by being beacons of His Light! *I'm preaching to myself first and foremost!

It is written: "This is the verdict: Light has come into the world, but people loved darkness instead of light because their deeds were evil. Everyone who does evil hates the light and will not come into the light for fear that their deeds will be exposed." John 3:20

Jesus is the true Light of the World. He said so himself and He proved it o'er and o'er. We must be joined to the Light… or else we will be exposed to the light. Do you understand… can you picture it?

We determined that by rubbing two sticks together you get fire. We talked about the heat… to exhaustion… to incineration… to ashes. Now let's talk about the other nature of fire. The light.

First, please allow me to share with you the other thing I saw in my three-day vision. While no two things could occupy the same space at the same time, I saw the contrapositive side of that, as well.

The cross is a paradox. The symbol of fusion and fission. It represents both death and life. Elimination and illumination. Not just from the act of Jesus' death and resurrection. No two things can occupy the same space at the same time… if it does it results in death. No matter how big or small. Our whole body can die from that Mack truck invading our space, or our ligament can die from a ski collision, or our skin can die from road rash, or cells can die under our skin, leaving a bruise after we hit our hand on the door frame. Ouch!

The cross is a symbol of death. However, the cross is also a symbol of life. Two things occupying one space. A union. I saw a pregnant woman which signifies two in one. They were sharing the same space. The womb made it possible. I also saw man and woman making love. Two becoming one. These unions represent love. It represents love in order to make life.

Two into one to create one again. A child is born. Instead of forcing it, like the creators of the hydrogen bomb (forcing two atoms to create a third), this process is natural and anointed with love, not death and destruction. A miracle takes place naturally. Love is transferred. Love is never by force. Force results in destruction. He will never force us to love Him. He will never rape us. He will never put His mighty weight upon us. Quite the opposite is true. If we have eyes to see, He will turn death into life. Jesus will remove the radioactive corrosive sin from our soul. His light cauterizes the spiritual wound. There is no shortage of testimonies... they're different but one thing's in common for all: those who lived a life of ick, then found Jesus, will tell us that it felt like a heavy weight had been lifted from them. They were carrying around, their wholes lives, the heavy load of crud and only the pure light of Jesus was able to burn away the "dross."

Jesus said, "Come unto me, all ye that labour and are heavy laden, and I will give you rest. Take my yoke upon you and learn of me; for I am meek and lowly in heart: and ye shall find rest unto your souls. For my yoke is easy, and my burden is light." Matthew 11:28-30

I am so grateful to God. As we lift up Jesus He removes the heaviness. Everything else we lift up (in worshipful admiration) we will eventually bear the weight of, except for Jesus. His *light* is light. His luminosity bears levity. There is no mass of heaviness to His countenance! He invited us to come to Him. To unify with Him in order to be like Him. How amazing is that! It is such sweet irony that in order for us to unload the weight of our burdens we must lift Him up! Our humility will open the door to the "cosmic carwash" for our spiritual vehicles. I have no way of becoming pure by myself. There is no amount of good deeds I can do to be clean enough... good enough. I can't do it on my own and there's no earthly soap strong enough (although Satan will tempt us with his serums, surgeries, and eventually, DNA modifications). No solvent can deter the ick from sticking and no detergent can remove it. Not permanently, anyway. We got one shot at this! One solution can dissolve all the ick. All the scum.

There's only one brand that can do the job! Everything else is generic and useless. Why waste our time? Nothing but the blood of Jesus (poured out from the cross) can clean our spiritual scum and heal our soul wounds! The promised Messiah. The HaMashiach, the Anointed One. The only one in history who was "expected" to come... the only one born with an RSVP to this life. And the only one who fulfilled every expectation. He is exclusive. One ticket per person (which includes a backstage pass). We can't ride the coattails of anyone else.

To enter His light is to be unified with Him one by one. Unlike Noah's Ark, where they entered two by two, the Ark of Jesus has a door custom fit for each of us. We all have our own entry. We want to do everything we can to find this gateway. Jesus said the narrow path will lead us there. We must find this passage. It is imperative we all find our way there before all gates are closed and the on-ramps removed.

His light purifies us, yes, but His light also protects us. Jesus is the light and His protective womb is a shield to the harmful radioactive substances on the outside of it. Inside is warm and full of love. If we are on the outside we are in the direct path of its exhaust. It's smoke. It's fire. We are out of the safe zone. It is hazardous to be outside of the light. If we are inside of the light when God flips the breaker switch it's like being in the life raft when the dam breaks or when the boat sinks. When it does, time will flow and if we are not in God's light we will be swept by its current and where it leads is not a good place, according to the Bible. I'm sure it is a downward spiral into the black hole of hell.

Jesus's light guides us and protects us, and His light cures us as it purifies us. Katie Souza, a meth addict and gun-toting drug dealer, was transformed by Jesus while serving time in Federal prison for her crimes. She had a supernatural experience which led to the discharge of her incarceration years before her appointed release date. She is now an evangelist focused on prison ministry. She believes in accelerated healing and transformation because, as she said herself, she only has a few hours to get inmates saved, healed, and delivered. Saved from sin, healed from affliction, and delivered from demons. I encourage you to listen to her own personal testimony. In her CD audio-set teaching called the Healing School she shares how she has overcome demonic kings by the King of Kings. She coined the term "light-laden DNA" when referring to Jesus's perfection and His ability to heal people. Katie is determined, just like Jesus did by his DNA-light-laden blood, to heal the sick physically and spiritually (from soul wounds) to live life victoriously day in and day out. By applying the blood of Jesus in prayer to our soul wounds first, we can have permanent restoration rather than temporary healing. Katie shared her own research and found that Dr. Vladimir Poponin, a quantum biologist at the Russian Academy of Science, proved that light is attracted to DNA. It's no wonder so many people today use light therapy for healing and is it any wonder that the light of Jesus is the light of all lights and offers the only permanent cure! I loved it when she said that God's light chases after our DNA to heal us because it was designed that way, hence the green rainbow around His Heavenly Throne as recorded in Revelation 4:3. It is glorious to behold the countless number of people all over the world finding this to be true. Jesus truly is the light of the world. He brings healing to all nations. He delivers us from evil, and he saves us from the torrent which seeks to suck us in to its vortex of despair, destruction, and death. Jesus is the captain of the ship which sails the seas of our lives; He is our rescuer, ready to throw us the life preserver to come aboard. All we have to do is realize we need rescuing and grab hold of His hand as He reaches out for ours.

We can't miss this boat! We cannot swim hard enough to escape the rip current. That's what it is. A rip tide. A 186 thousand plus miles per second tidal motion... the speed at which this stream flows in the river of time without God's *space* will rip us into a sea of abandonment. Too many of us are on a party boat not aware of the fact that we are circling the toilet bowl. We cannot believe Satan's lies. He is not God. We are not God.

We will not be able to "positively" think our way out of it. We do not have the power to remove the ick (as it is) by ourselves even if we meditated day and night for the rest of our lives. Nor will the ick go away if we ignore it. We need Jesus, but He's not for sale. He is a gift! The admission ticket to enter His gate is free. None of us deserve it but He loves us anyway and desires our company! What more could we possibly ask for? Regardless of our own awareness we will all perish without Him because he is the light and life source.

Romans 2:23 says that we *all* have sinned and all have fallen short of the glory of God. Sin. There's that word again. Makes people shudder. Sin is the same thing as ick. Now maybe after publishing this book, it will be taboo to use the word ICK too. Satan's motive to deceive us and cause us to sin is fueled by hatred and jealousy. God loves us. We are His children. We have the birthright and the inheritance. The devil passionately hates us for it. Satan is the liar pointing his lying finger at God, Jesus, and the Bible. What would motivate the writers of the Bible if it were all just a big lie? If Jesus were a liar what would motivate Him? Think about it. Believing in an all-loving God has no underlying ulterior motive. Sure, a snake in the grass acting like a loving preacher could take advantage of a crowd. That's possible and unfortunately, a real scenario in many churches. Yet, it does not negate the permanence or power of the written or Living word. One must know how to differentiate between unmoving solidarity and quicksand. Coincidentally, the Word is the meter by which to make that determination.

Satan is out to deceive us in order to rip us from the one who made us and loves us. If the devil gets his claws in us we are attached (connected, for real) to him (the ICK of all ICKS) even into the afterlife. We need an atomic knife to sever the tie before we draw our last breath, and according to scripture, we have been given the Sword of the Spirit of God to do exactly that!

Scripture. That word has become taboo too; and so has the Bible. Here's the thing, I have read a lot of books, but the Bible trumps them all. There is no comparison. There is a good reason why the Bible was the very first book to be published on the Guttenberg press in 1454 AD. Before that, the Roman Catholic Church was the only one who had copies. The Papacy would allow only one Bible per church, usually per town or region, not just because it took so long to handwrite copies, but because they wanted to keep the knowledge for themselves and not allow common man to have such knowledge (that's why Luther, and other Catholic (monks) left the Roman church to *protest*; thus began their own *protest*ant teachings). Once the Bible hit the press, everyone was able to have a copy, the Roman church began to lose power. God loves us so much He wants all of us to know His secret place, not just the "heads of state."

The Bible is still controversial. Through the centuries, many argued over translations, dividing the once united protestant followers. King James released his translation in 1611 and Miles Coverdale released the Geneva Bible in 1557 A.D., both of which contained the apocryphal books (the books not in the modern Biblical canon), but since then has been removed, yet the Roman church maintained the apocrypha in their copies. The apocrypha (at least some of them) were in the original Septuagint (translation of the Hebrew into Greek, circa 132 BC), so who says they were not inspired by God? Who authorized the canon we use today? The protestants? The Romans? The church of Laodicea? And how can we trust we have the whole Bible today? Did God intend it that way or did someone succeed in the violation of Revelation 22:18? I question this only because the "approved" Biblical Canon we use today makes references to Enoch (an apocryphal book). Jude refers to Enoch as a prophet of God, which would make *Enoch's book* God's prophetical word, would it not? Jude quoted him (from Enoch's book) when referring to angels leaving their first estate (Jude 1:6). Enoch was highly favored as he did not see death, according to Genesis 5, so he would qualify, surely, as one of God's prophets of whom we should all take heed, right? The Book of Enoch supports the Bible as we know it and further confirms a coming Messiah, so I recommend it and do not see anything blasphemous about it. Enoch successfully answers the Genesis six question about the fallen angels and satisfies all our questions about angels, demons, demigods, giants, chimeras, and ancient aliens. It is Satan's pleasure to dismantle the Word of God anyway he can in order to keep the truth from us, and though he tries, he cannot succeed. The truth always rises to the surface.

The Bible. It's Satan's and man's fault that it has become a detestable book to modern culture. Don't be fooled. The Bible is the most important book we could ever have the privilege of reading. More people have died for this book than any other book in history, and we want to "shelf" it? Get it and read it before Satan succeeds in burning all copies (like Hitler tried to do).

If anyone tells you the Bible cannot be trusted for whatever reason, they are ill informed or their hate for Jesus clouds their reasoning. In all its translations, and *as is*, the Bible has been proven time and again to be the most reliable collection of documents the world over. Side by side when testing the bibliographical texts with other historical documents such as the Iliad, or the works of Herodotus, the Bible is the winner by a landslide. The winner to ALL of them. Don't take my word for it, check it out for yourselves. No ancient text can compare, and the measurable contrasts are laughable. If anyone tells you differently, they have not observed the striking truth for themselves or they choose to be in opposition regardless of proof (as in the same proof-positive tests we use in a court of law to determine valid testimony and acceptable evidence). In addition, all the Bible prophecies which have come true reveal unimaginable odds and prove it as being super-duper-natural. The fulfilled prophecies of Jesus alone (much less the rest) have odds that are out of this world, literally bigger than our universe... the mathematical probability for just a few of the hundreds is broken down here from Bible Timelines website:

There are over 300 prophecies listed below that point directly to the Messiah. Here is an example of just 8 : The time of His birth (see the Daniel 8 & 9 Timeline). He would be born in Bethlehem. (Micah 5:2) He would be born of a virgin. (Isaiah 7:14) He would be betrayed for 30 pieces of silver. (Zechariah 11:12) He would be mocked. (Psalm 22:7,8) He would be crucified. (John 3:14) He would be pierced. (Psalms 22:16) He would die with the wicked, but He would be buried with the rich. (Isaiah 53:9)

Mathematics & Astronomy Professor Peter W. Stoner has made the statement that the chances of just 8 prophecies (like these) coming true by sheer chance is 1 in 1017 (100,000,000,000,000,000). That would be equivalent to covering the whole state of Texas with silver dollars two feet deep and then expecting a blindfolded man to walk across the state and on the very first try find the ONE coin you marked (roughly equivalent to the Province of Ontario being 1.5 feet deep). And if we were to add only 8 more similar prophecies, for a total of 16, the odds would be 1 x 1028 x 1017 or 1 in 1045 (1,000,000,000,000,000,000,000,000,000,000,000,000,00 0,000,000) Using the same type of illustration as above, if we were to press this many silver dollars (1045) into a ball and place the center of this ball where the center of our Sun is, this silver ball's outer edge would be in approximately the same area as the orbit of Neptune (almost to Pluto.). One man fulfilling all 16 prophecies by sheer chance would be like sending a blind-folded person out to find one specific silver dollar that has been marked and has been mixed up somewhere in this huge ball and actually finding it the first time! (Keep in mind that this is a three-dimensional BALL, not a disk like our Solar System.)

Professor Stoner gives us yet another illustration, but this time, because a silver dollar, and even the atom, would be too large for this one he chose to use just the electron that orbits the nucleus of the atom. The electron is one of the smallest particles of matter known to man. It is so small that if you lined up 2.5 x 1015 (2,500,000,000,000,000) single file you would end up with a line that is only 1 inch long. (If we were going to count the electrons in this line one inch long, and counted 250 each minute, and if we counted day and night, it would take us 19,000,000 years to count just the one-inch line of electrons.) Now back to Stoner's illustration. Using only 48 of the prophecies that Jesus fulfilled it was calculated that the chances of one person fulfilling these by sheer chance would be 1 in 10157. To help us understand how huge this number is he suggested taking 10157 electrons and pressing them into a solid ball. This ball, made entirely of electrons, would pretty much fill, not just our galaxy, but the entire known universe. (At the time Stoner's book was written the universe was known to be at least 6 billion light years in all directions. Stoner's known universe would need to be filled about 10,000,000, 000,000,000,000,000,000,000 times.) Once again we would mark just one electron, blindfold a man and send him out to find that one electron. Peter Stoner then states, "To the extent, then, that we know this blindfolded man cannot pick out the marked electron, we know that the Bible is inspired. This is not merely evidence. It is proof of the Bible's inspiration by God--

proof so definite that the universe is not large enough to hold the evidence."

Professor Stoner then made the comment that, "Any man who rejects Christ as the Son of God is rejecting a fact proved perhaps more absolutely than any other fact in the world." (For more details on these illustrations and others by Professor Peter W. Stoner, you can read just his chapter entitled, The Christ of Prophecy, or you can read his entire book, Science Speaks, online.) And the fact that Jesus had no control over things like where, when and how He was born, or where He was buried just adds to the available evidence that He did not just work to make sure that all of the prophecies were fulfilled so that people would believe Him to be the Messiah.

http://www.bibletimelines.net/article/24/articles-brief-and-to-thepoint/jesus-is-he-really-the-messiah

To see an entire list of prophecies fulfilled by Jesus check out

Mathematical+Proof+for+the+Existence+of+Jesus+by+Prophecies+Fulfilled.pdf

Believing in the Bible is an intelligent choice. We are to love God with all our mind, the Bible says. God made it easy to do for children and adults. It's another wonderful paradox. It is simple to understand the main messages, yet there are enough mysteries to intrigue the most inquisitive minds. For the slow learner, the savant, and the genius. There's something for everyone and enough proof to go around! And around. And around. Forever. The scientific method used to determine the astronomical odds (for only eight of 300 total fulfilled Bible prophecies, mind you) should be proof enough for anyone who knows how to count to ten. Yet, God didn't stop there… the Bible also has codes within it which continue to increase the odds to bigger than unfathomable proportions. Equidistant letter sequencing (ELS) is found in the Bible. Some may tell you codes can be found randomly in any large text, but they're overlooking the obvious. You may find one or two, a handful at most, as was tested using very large text-sized books such as War and Peace and Moby Dick, but nothing like the overwhelming Bible codes.

The code in the Bible was probably the most amazing and significant discovery in recent history. In all of history, for that matter (especially if it is the opened seal of Daniel's prophecy), yet it was bypassed as if unimportant and barely made headlines. It is not surprising that the secular world would dismiss this incredible find as banal, but it is remarkable that it dissipated into fumes from the minds of so many Bible believers. Fumes. Here, we have the most incredible, solid, evidence that the Bible was not written by any earthly intelligence, but I rarely, if ever, hear it used to defend the validity of the Bible as God breathed. It was proven, mathematically, that all the manmade computers combined could not have duplicated it. This discovery should have catapulted the Bible to gain mainstream respect. Now, more than ever, the Bible is reduced to a questionable book among the masses, when it should be raised to an unmatched status of historical text.

The internet is inundated with sites debunking the Bible code. Their arguments are unimpressive, especially compared to the codes themselves. They minimize the experiments of Israeli mathematician, Dr. Eliyahu Rips (one of the founders of the code) and physicist Doron Witztum. Critics tout books like Moby Dick and War and Peace as encoded, therefore, they say the Bible codes are not unique. They fail to show a side-by-side comparison of those books' coded records compared to the Bible. If they did, then it would prove them wrong. They make claims but never once mention the fact that the Bible's code had been verified by peer review studies (expert mathematicians of quantum mechanic's field study of group theory) who did compare the Bible to

Moby Dick and War And Peace and other books to determine its validity. They fail to mention a senior code breaker from the NSA, Harold Gans, who set out to debunk it instead became a Bible believer after more than 400 hours of testing proved it to be true!

When I first researched the codes in 1997, there were only eight sites on-line, now there are 196,000,000 results to my search. The waters are muddied... beware of broad falsified "facts" and misinformation disparaging the codes. Ironically, in Daniel chapter 12 when the archangel Michael tells Daniel to seal up the scroll until the time of the end, the coded part may imply the code itself as the "secret" sealed scroll to be opened in the *end times*. The Bible Code (the book published in 1997) discusses the probability of the computer age as being the "time of the end" timeline of this oracle. Did you know Isaac Newton spent the latter years of his life searching for codes in the Bible? They don't teach you that in science class do they? Newton believed all of creation foretold in the Bible was a cryptogram and this Daniel 12 prophecy alluded to it. By comparison, most of Newton's journaling detailed this research and not so much on the laws of gravity and motion. Allegedly, he was obsessively certain the Bible contained hidden messages but, unfortunately, hard pressed to find them. He was right. There are hidden codes, but it would take another 300 years for a computer to find them and Hebrew scientists to break them, but not before Jewish Rabbi, Michael Dov Weissmandl, who was known for saving Slovakian Jews from concentration camps in WWII, became the first man to discover the first set of ELS codes within the Bible, known as the Torah Codes.

Skeptics are everywhere, like cucarachas! The code is additional undeniable proof, yet people {scholars and professionals} will still turn their heads and scoff. They hate Jesus so much they deny mathematical proof! Blows my mind! There's a lot of literature published about the codes, but I recommend Michael Drosnin's book, The Bible Code, which was the original mass debut of the codes back in the 90's. Drosnin was a journalist for the Washington Post, not religious... his book is not theologically based nor is it a literary gem, but his research was impressive. He went to Israel chasing a story and came back a believer and had more information than would fit in a newspaper column. He published his findings in a book. I highly recommend it. Non-random ELS codes are exclusive to the Bible despite critic's claims. The Bible is unique because ELS sequences are abundant in the coded matrix of its text which was proven not to be random. I didn't take anyone's word for it and you shouldn't take mine. Research it yourself. Start with Drosnin's book which details fulfilled coded prophecies; he has included Hebrew copies of the Torah charted with codes so you can see for yourself how ELS works. What's more, since then I have read an article claiming it was found that some codes create mosaic images. Amazing! The odds are infinite just like God Himself. He said He is the alpha and omega, the beginning and the end. It proves the Bible was not written without the inspiration of something "extra-terrestrial" or rather beyond the brainpower of mere man. The next time you hear someone say you cannot trust the Bible because it was written by man, you can be sure they're ignorant to the truth. Furthermore, many high-profile critics of the Bible hold a double-standard. If "spirit writing" is documented as a valid occult practice (which is becoming more popular and culturally acceptable), why wouldn't God's Holy Spirit move the hand of man? That's hypocritical logic, wouldn't you say? I don't know about you, but I'd rather place my faith in Holy Spirit inspired writing than anything else. If it is not Holy Spirit-inspired then it is less than holy, right? God made it so the proof is undeniable so in the end no one is without excuse. I have a lot to say about the codes as I have researched them thoroughly... wish I had more time, but I will say this to wrap this up... if the codes prove the Bible was written supernaturally (and, btw, codes were also found in the Peshitta Aramaic New Testament text) couldn't we test the apocryphal books the same way?

For more info, check out:
http://ad2004.com/Biblecodes/Greekmatrix/Grkmatrix.html

I was once entranced by horoscopes, numerology, and gematria but since then I have refrained from participating in those things. I do not want to peek into the future. Reading about the Bible Code is one thing but seeking the Bible for codes in regard to our future is another (besides, I'm satisfied from the Bible's un-coded (plain) text and I get more than enough forecasting from apocalyptic visions and dreams). I have no interest in the Bible codes for that manner. I'll say it again for emphasis: The Bible warns against soothsayers, fortune tellers etc. because of their potential use as a medium by Satan to lure starstruck and unsuspecting people into his trap by forecasting future events. This includes human mediums as well as objects, astrological horoscopes, and ciphers. There are many oracles... a dime a dozen these days. I realize God is a numbers guy and that is why numerology and the zodiac intrigued me in the first place. True, dates and numbers are significant to God, but I personally, will not decipher them. I'd rather stay safe than sorry. I've seen too many times, Bible-believing Christians fall into the trap of date-setting because they were so convinced by the numbers from which they swore were God-inspired to be true, only to be disappointed when their prediction did not come to pass. This practice is forbidden by the Bible for good reason and it doesn't serve well as a testimony when forecasts fail to fulfill future events. As far as *when* Jesus will return, we should be ready *today* every day. In addition, we must listen for the trumpet calls by paying attention to the Holy Spirit which points to the Hebraic calendar, events, and God's appointed watchmen who are placed on the spiritual wall of today. Who are they? The Bible says they are sons and daughters, young and old men... people from all nations... see Acts 2:17 and Joel 2:28.

The most common argument among the church in this modern era is in regard to the 2nd coming of Christ. Pre-tribulation versus post tribulation rapture... or mid-trib seems to be the hot topic. I discuss this subject in book three. I will not get into it here.

After that first year in and out of the Bozeman library, I reached out to "professionals" because once I had the language and understanding of particle physics I longed to have deep meaningful conversations with someone… anyone. I reached out to many. No one reached back. No one except one. His name was Mac. Mac Rugheimer. Dr. Rugheimer was a tenured professor of Physics at MSU. He became my buddy. I began attending a Jewish Messianic Church with him. We had similar passions. He was a lifelong atheist, teaching to all his students that God didn't exist, until just a few years before I had the pleasure to know him. You see, he found a macro code in Genesis 5, in the genealogy of Adam to Noah. It changed his life. The Genesis 5 Code changed Mac's preconceived atheistic worldview and made him a Bible believer. The odds just for that one set of code astounded him (he knew how to count to ten, lol). In fact, he gave a special lecture (post-tenure) at the University revealing these findings back in the late 90's, and I was one of the lucky attendees! He discovered that all the names in Genesis 5 revealed the gospel of Jesus. As you know, the Hebrew alphabet makes up their numerical system and every word and number have specific definitions, as well. Here are the name translations/transliterations of Genesis five:

Hebrew	English
Adam	Man
Seth	Appointed
Enosh	Mortal
Kenan	Sorrow
Mahalalel	The Blessed God
Jared	Shall come down
Enoch	Teaching
Methuselah	His death shall bring
Lamech	The despairing
Noah	Rest and comfort

62. Genesis 5 Genealogy Record/Translation

Figure 62 displays the genealogy record. The left column shows the Hebrew names and the right displays the names translated into English. By reading the meaning in order, Mac was convinced this was a code of the coming HaMashiach which Jesus had fulfilled. *Since then, the Genesis 5 code has hit mainstream and has become very popular (you may have seen it or read about it).

Mac chose to receive the invitation Jesus made to all of us. In that lecture hall at MSU Mac blew the trumpet of God, declaring His living word and giving glory to Jesus! Wow. Dr. Rugheimer was brave and risked his long-standing reputation as a scientist to honor Jesus. He was very much respected, and not just in his school. He was the first physicist west of the Mississippi to build a holography lab many years before, earning Montana State University an honorary place in the academic community of science. Although he was retired he still spent many days at the school volunteering. He has since passed away but there is a physics scholarship fund in his name at MSU to memorialize him. He is with Jesus now and I look forward to seeing him again one day. The Bible is trustworthy.

Whether we have the entire Bible in the Canon today, the protestant Bible, as is, and in all its many plain text translations is still legitimate and furthers the probability factor to balloon even bigger. This has been proven by many scholars, but I recommend The Case for Christ by Lee Strobel. It is a book (there is a movie too, available on Netflix) about another atheist; this one hellbent on disproving the validity of the Bible by disproving Jesus died and resurrected. It is very revealing. His book addresses the criticism most people have about the various translations of the Bible making it untrustworthy. His findings proved the opposite to be true. So, breathe easy, whether it is the King James version or the NIV, the main message of salvation is the same, and that's what matter most. If you would like to know how variations of translations and versions help to validate testimonies then I urge you to read Lee's book! Strobel, just like the Apostle Paul who was hellbent on persecuting Christians in the 1st century, ended up becoming a disciple of Christ instead. These men, two thousand years apart, hated Jesus so much... what changed their minds? Look it up! There are numerous testimonies just like theirs within the two thousand years between them. See for yourself. Your soul wants to know.

Furthermore, the record keeping practices of the Jewish people for thousands of years were unmatched. Their attention to detail is superior. The Dead Sea Scrolls (excavated from Qumran in 1946) prove this statement to be true and further proves the historical texts within Bible are the same today as they were when the Essenes scribed them thousands of years ago. Also, Yemenite Jews were scrupulous scribes, having only nine varying letters from more than 400,000 differ from the Masoretic text. It was confirmed the nine errors did not alter the meaning of the message. This was all before hitting Gutenberg's press for mass publication. I've read arguments about the Hebrew and Jewish scripts having some problems. Not so much Hebrew to Jewish, and not so much from Aramaic to Jewish, but from Moabite (a Canaanite dialect) to Hebrew. If you would like to read more, visit http://www.talkreason.org/articles/list.cfm

Filling in the blanks with other texts can be useful as long as it doesn't take away from the overall message. If we read a text that does, it should be discarded because it was not written by the Holy Spirit. The goal of the anti-holy spirits is to contradict the Holy Spirit, so keep that in mind when taking into account all the various texts out there. As far as I'm concerned, an all-powerful God can protect and preserve His word easily, and if He intended for us to receive His message, no doubt, no one or no thing can stand in His way!

The more translations and versions the better. Makes the hunt for truth all the more challenging and fun. I am impressed with new discoveries of Aramaic, Hebrew, Phoenician, etc., translations of our modern texts. It has not dulled my faith, oh no, it sharpens it because it supports the Bible as we know it to be true. Sure, there are nuggets to encourage more study and examination in regard to the mysteries of this and that, but it does not dilute the truth of Jesus as the savior. The old testament is a testament to Jesus the Messiah. It scripted the markers and "address" of Jesus to a tee. It's true, varying translations as well as varying interpretations lead to differing denominations of the Christian church. In my walk, it has become increasingly obvious that the main message and Jesus being central to it all is clear and most important. To take away from the simple message of salvation and the sovereignty of Jesus one has to do a lot of unnecessary contextual adding and omitting. If we read the text rather than someone's interpretation of it we can see the truth for ourselves. The supernatural movement of the Holy Spirit comes from reading the Bible directly, not man's piecemeal version of it. Only until we read it ourselves will we see there are no provisions required.

Moreover, the fact that the Jews are excellent scribes and can maintain their manuscripts meticulously is second to the fact that the Jews themselves, as a nation of people, are evidence of their history. They are the longest running nation still thriving today and their gene pool can be traced all the way back to Adam with their genealogy record. And then it just ends. Think about that. It just stops at zero counting backwards. It didn't just stop because it was interrupted by spontaneous evolution. Perfect record keeping had a beginning. KISS. Bottom line: The Holy Bible can be trusted. Truly, it is the only book that can be. The mysteries are endless. The teachings are timeless. The love of God is palpable. But so is His strong arm. His wrath for the ick is real and we would be wise to fear Him.

For those in the light versus those in darkness: "Therefore rejoice, you heavens and you who dwell in them! But woe to the earth and the sea, because the devil has gone down to you! He is filled with fury, because he knows that his time is short." Revelations 12:12

There will come a day when God will withdraw His light from the world. There will come a day when His wrath will come upon the wicked. Woe to you if you are not in the light. Even nuclear bombs are not to be feared if you are in the light. Jesus said fear nothing except God himself, "Do not be afraid of those who kill the body but cannot kill the soul. Rather, be afraid of the One who can destroy both soul and body in hell." Matthew 10:28

Hell. Another no-no. Too many taboos. Hell is a big subject (and a big place, we're told) and I only skimmed the surface (hellfire would leave no trace of the surface). It is for another book, no room or time here.

Those in Heaven with God are in the light. But so are we on earth if *Heaven dwells in us*. Jesus gave us The Lord's Prayer, so we know what to pray for: "Our Father which art in heaven, Hallowed be thy name. Thy kingdom come, *thy will be done in earth, as it is in heaven*. Give us this day our daily bread. And forgive us our debts, as we forgive our debtors. And lead us not into temptation but deliver us from evil: For thine is the kingdom, and the power, and the glory, forever. Amen."

"Thy Kingdom come... in *earth* as it is in Heaven..." Jesus said it is God's will for Heaven to come down and we ought to pray for this all the time... those who pray this prayer and "walk" with the Lord, have Heaven come down into their hearts, and are *in the light*. This *is* the will of God. This is what our Heavenly father wants for all of us in this dark world. You want to be in the light? Have Heaven come down into you! Do it today! Today is the day of salvation! Time is running out and if you are not in the light when God flips the switch... heaven help you. The thing is, when that happens, Heaven won't help you. God said the flipped switch means game over.

What do we believe and why do we believe it?

When it comes to our eternal soul, do we really want to pass the buck to some "professional?" An "expert" in spiritual matters? A therapist. A priest? A guru's teaching which the Bible warns is idolatry? Isn't this the one area where we want to do our own research, just to make sure? Do our own math? Don't believe "it" just because we were "raised that way," either. Our parents do not have the final word. We owe it to ourselves to find the truth by ourselves. It should be everyone's duty. Not a hand-me-down.

I had a biology professor for whom I completed my work study at Western State during my junior and senior years. I looked up to him tremendously until one day I had shown him an over-large coffee-table-book my boyfriend had given me as a Valentine gift the day before. It was a glorious picture book of sketched animals in nature joined with the theme of Noah's Ark. Proudly, I thumbed through the enormous pages, admiring the sketches and making small talk of how Noah's ark was my favorite Bible story. To my dismay, he ridiculed me, "You don't actually believe in that "bologny!""

With a spirit run over and a forced smile I replied with as much gusto as I could muster over a croaked voice, "Why not?" Little did he know that he lost a fair amount of favor and respect from me that day. Not because he had a differing viewpoint, but because he couldn't give me a satisfying reason why it was bologna as he belittled me. He didn't present me with any reliable data unless you call the theory of evolution reliable.

I continued working as his assistant, grading papers, making copies, but mostly drawing anatomical figures of pre-historic fishes and reptiles, and simple-celled organisms as well as human internal systems for the following semester's Histology lab manual. Cooperatively we worked side by side after classes, before classes, and in between classes, all the while something inside me was brewing. His outspoken disdain for the Bible was eating me inside and I wanted to know more about evolution versus creation.

Instead of rebutting with holier-than-thou adages and New Testament Gospel, I quietly read to myself books he held near and dear. Authors of the "gospel" of mass evolution. From Charles Darwin himself to Stephen J Gould, Jared Diamond, and other notable anti-creationists, I studied their theories, their research, their findings. Little did I know at the time that I was well on my way to becoming an unofficial scholar on the subject.

One day, I awoke and found myself debating creation vs evolution with the best of them. In fact, I joined the school's Forensics Team just so I could debate evolution in a "real" debate forum, but I never got the chance (oh, and (at the time) to debate the benefits of legalizing marijuana, lol). The Team assigned the subjects to be debated in regional symposiums, so I had to stick to the assignments. I was never given the opportunity to debate either subject. It defeated the purpose of why I joined in the first place, so I dropped out. Nonetheless, I knew nearly everything there was to know about the theory of evolution. I went all in when I studied it. I left no prehistoric rock unturned. From Neanderthal and Cro-Magnon to Piltdown Man and Australopithecus, from the big bang to the primordial soup. From the building blocks of protein to natural selection. None of it held water, especially not the soup from which *we single-celled organisms evolved*, and (theoretically) crawled out of with *our amphibious appendages... then reptiles, birds... monkeys... man.* They're missing the *missing links*. Lol! I couldn't get over the hypothesis that a primordial soup of our oceans would aid in polymerization of atoms, much less amino acids. I could not reconcile that in my pea-brain, so how did the "mega brains" of science do it? How do our cells stick together? How do our atoms connect if not for a thread? And if superstring really could weave us together who was the seamstress? Where did the needle come from? *How come no one seems to care that a maker makes more sense?* I cried.

Also, I found the history of this teaching of evolution was full of make believe. Turned out Piltdown Man was forged from the skull of a pig by an atheist hellbent on disproving creation. Piltdown Man's discovery made headlines. Proof that it was bunk did not. Go figure. Satan controlled the press. He still does! I found more fakes from the proponents of Evolution. It was pathetic. You must have more "blind" faith to believe in Darwinian Evolution than the fake news that reports it as fact! Speaking of which, if you look at a publication on evolution from like the 70's or 80's and compare it to another article published years later, you will notice time period changes. I believe they increase the years for time periods (ie, Jurassic, Paleolithic, etc.) to compensate for the odds against it. Clever, but can't fool me. Fyi, the variables they inject into their equations to arrive at those dates are all unknown. UNKNOWN. Think about it. They find a fossil and tell you it's millions of years old. Gimme a break. The Earth strains to preserve fossils 4,000 years old. Even the half-life measurements of carbon (^{14}C) don't go that far back in time (*and do not believe the Darwinian supporters who tell us the half-life never reaches zero) Everything reaches zero (the womb of the cross)... that's where it all comes from, lol! Let me ask you an honest question... wouldn't you think any remains of earth millions of years old would be so weathered with age that it would be unrecognizable? like oil... a *fossil fuel*? And btw, why does our government confiscate fossil evidence that proves earth isn't as old as they claim it to be or re-imagine geological evidence that contradicts their manmade narrative (like a world-wide flood and a race of giants)? Supporters of evolution continually modify their theses to support the Darwinian narrative to counteract arguments against it. The Bible and its stories have remained constant while the Theory of Evolution evolves with every new harebrain discovered. I'm told I need to read one of Richard Dawkins' books, *The Extended Phenotype* or *The Selfish Gene*. *He is no harebrain!* Dawkins seems to be hailed as the "prime-minister" of evolutionary teaching. Perhaps I will pick up one of his books because using the human genome as "proof" of biological evolution piques my interest. I am confident our genes are evidence to the contrary and I have no doubt I will find perforations to any evolutionary "proof" which claims otherwise.

This whole thing with my biology professor and Noah's Ark spawned in me a desire to know the truth beyond Darwin. I wanted to learn the faith of others outside of the Bible stories I knew. I wanted to hear their stories from their religions. I had to be fair. I'm a reasonable person. So, I made a list. And one by one I eventually crossed every religion off my list. The process of elimination is an effective tool. There were no close calls. All religions failed miserably for more than one reason. The Bible kept shining brighter and brighter the whole time. For real. Faith is a gift from God, and even though I cannot see or touch the stories I believe in, they were becoming more legitimate as I viewed them indiscriminately side by side with other stories. In fact, it brought the other stories deeper meaning. How amazing is that! The word in the Bible really was *living*. The same verses I had read a hundred times had new dimensions of revelation. This caused my faith to grow by leaps and bounds and this began my unwavering determination to search out the truth in everything. Not just the *Who Am I's. Who Are You*?

What's more, the story of Noah's Ark and the Flood is backed up by so much proof... the archaeological evidence for all Bible stories is overwhelming, but this one Bible story alone has undeniable evidence on Mount Ararat in Turkey. Never mind the fact that nearly every nation on earth has a grand flood story in their ancient history and marine life has been discovered on mountain tops and in barren deserts far from the ocean, the ark has real fossil record to support it; more than I can say for all the records Evolution claims to have for its entire theory. Don't believe me? Good! Do your own research, but don't be biased about what you read. Read it all! Satan's lies are everywhere. I'm sad to learn that Darwinian Evolution is now being taught as fact in schools today. Another point scored for Team Satan! I hear teachers and professors belittle and muzzle students for offering creation as a rebuttal to their opinion. Science is being contorted and taught religiously as another form of resistance to Jesus. Ironic, isn't it? *Jesus is too religious.* But not humanism!

There is no shortage of arguments against Jesus and/or the Bible. One by one I examined them. Controversial disagreements range from the origin of species to the origin of texts. Many argue that Sumerian cuneiform tablets prove the Bible to be unreliable because the tablets pre-date the writing of Moses's Torah. I have a couple things to say about the validity of this dispute. First, the content of the Torah predates Mesopotamian civilizations and even provides a genealogical record of their existence which substantiates it. Secondly, the ones who make this argument, hypocritically, accept theoretical accounts made by modern (last couple hundred years) men (ie, Darwin, Huxley, Dawkins, etc.) and place more faith in them than in the Bible. Hypothetically, if the Bible were written yesterday, but still had the same mathematical odds of being supernatural, it would still trump all other texts. Its timeline is irrelevant if God indeed wrote it and proved it with math and archaeological evidence. God used the hand of man to put pen to paper, and it just so happens those men (prophets, servants and Kings), who lived hundreds and thousands of years apart from each other were congruent in their writing and carried the same signature of the Hebrew God we know of as Elohim. Adonai. Yahweh. El Shaddai. Jehovah Jireh. YAHOVAH. YHVH. I AM. I am that I am! The God of Abraham, the father of the Hebrew nation, proved Himself. The Hebrew carpenter, Yeshua HaMashiach (Jesus Christ) did so, as well. There is plenty of evidence that Jesus claimed to be God, (not just *a god*). That alone is no big deal. Anyone can claim to be God. However, Jesus proved it. Don't take my word for it. Research it with the same non-bias examination we use for all historical documents. Those who have already (including many atheists set out to disprove the divinity of Jesus) are now riding the Jesus train! All over, people are waking up to the realization that the Bible can be trusted and should be trusted, myself included.

And so we have the prophetic word confirmed, which you do well to heed as a light that shines in a dark place, until the day dawns and the morning star rises in your hearts; knowing this first, that no prophecy of Scripture is of any private interpretation, for prophecy never came by the will of man, but holy men of God spoke as they were moved by the Holy Spirit. 2 Peter 1:19-21

I did not want to be careless with this. It deserved the deepest introspection I could give it. The Bible. God. Faith. Jesus. The alternatives. Although I knew in my heart God was my Abba (father), I had to know in my mind if it was intelligent. I chose not to take anyone's word for it, especially not my mother's. Priests, professors, gurus... it didn't matter to me how enlightened or moral they were, I would take the wheel on this one. I didn't care if they had 10 PhD's and 20 Nobel prizes, I would be the judge. A rocket scientist or a supernatural healer made no difference to me; I would check it out for myself. No human was qualified to make this decision on my behalf. *My soul is at stake.*

I had lots of questions. Still do. As children, our first questions to the WHO AM I's are met with answers from our parents who, most of the time, had the same questions they did and the same questions their parents did. They give a generational answer. Oma, having been raised in a predominantly Buddhist nation, did not give me the same answers her father gave her. Truthfully, I do not recall having this conversation with her or with my dad because they let Bible school fill in the blanks. That was fine with me. Parent today, Bible believers or not, should allow their kids to attend Vacation Bible School at least once, just to give them the human right to explore all answers to their inquisitive minds. It's only fair.

From an early age I carried a peace with me that instinctively knew God was good, the devil was bad, and Jesus was... well, uh, Jesus was my guardian angel. I know now that He is *more* than an angel, but back then, that's how I perceived it. As a child it gave me peace of mind and quelled my nightmares. I was fortunate to have had Jesus with me through struggles and dark encounters because He protected me. Whether it was His appointed angels or not, I knew the protection was from Jesus one way or the other. I never doubted it. In spite of my parent's shortcomings, I appreciate the fact that I was taught both sides of the story when it came to the Bible. Heaven *and hell.* I learned not only about the wonderful side of the story... Noah saving the animals and preserving their species... and of the dove and the olive branch. But I also learned why God destroyed the world with a flood. I was not pampered. I got to see both sides. This very much helped me become the woman I am today. Even in all my bad choices growing up I was never disillusioned by obscure ideas of morality or ethical do's and don'ts. Good from evil or vice versa.

When I would find myself suffering from a hangover or the sticky residue of my own making I knew it was my fault. I realize many of us walk a path away from God, away from Jesus, because something happened in our lives, usually as children, that caused us to reject Him, whether it was from an unanswered prayer or a person(s) who led us to believe God was not real, or He didn't care. Still, many of us never had the opportunity to know Him at all because we were raised in other belief systems. I've noticed, however, those of us who reject [the Biblical] God did/do so because of resentment, anger, or apathy. And those who had not known Him from childhood yet found Him, end up accepting the Biblical account with over-the-top gladness, as if they found hidden treasure. Their encounter is more intense than those who knew Him all along. All nationalities share the same joy.

Those who reject Jesus/God/the Bible because they feel there is no scientific proof, I've found, resent Christianity in spite of the plethora of proof. They rather believe what Confucius says than have any association with Christian values. It's the taboo mentality Hollywood has so successfully ingrained in everyone. If "proof" was all anyone really wanted, then the scientific method of probability disclosed above would be enough. How much more proof does anyone need? The Bible is clearly the most *probable* of truths given all possible choices. The math is undeniable. Yet, most deny Biblical proof and choose to believe it is not real.

Jesus was always real to me because anytime I was alone with myself, feeling sorry for myself, He was there to relieve me. Whether it was in my own head (as sceptics would say) didn't matter, it was a real relief, a real warmth, a real tenderness, a real friend. *Fyi, modern science is beginning to take notice and legitimize the *realness of things in the head* as real things. After all, information is the coin of the realm, and our minds are the registers. I had an instinct early on, unlike most, that things in the head were just as real, if not more, than things I could see and touch. I never blamed God for bad things. I had the presence of mind to know that He was sad about bad things just as much as I was... more so. During the consequences of all my bad decisions, accidental injuries, or "natural" sickness, no one could convince me that Jesus was not real. Even if I couldn't feel His presence and even if I were in the middle of a temper tantrum, I knew He was real, and He would heal me, somehow, even if I didn't deserve it. This was before I did any research. Studying history, science, and religion answered my questions, but not to convince me Jesus was real. I studied those things to see if the real Jesus (I knew in my heart) had any scientific or historical proof in the world at large. Yes, what I found fueled my faith. Not only was He real *to me*. He was the real deal, and everything else was conjured. I figured this out on my own accord in a fair assessment of all accounts I could find in recorded history and science. I did my own research.

We put too much trust in other people. We throw them the ball then sit on the bench. We let them take the wheel on roads where we should be driving. We are too busy to do the math ourselves, so we have a professor... a "professional" or a parent do it for us, and we tend to believe what they say as gospel because they are the pros... or father knows best!

Why do we consider ourselves enlightened by other's "bright" ideas without first checking their source? The real origin of it. There could be a snake hiding below the surface. The snake who twisted God's truth into a crooked beam of light and who took the soot of the smoke from the fire, spit on it and fashioned it to look like the real deal. What idol of his carving are we cherishing? *We set it on a pedestal under the spotlight and applaud. The light makes it look pure. It gives it a glow.* It's all smoke and mirrors. We foolishly fail to look beyond the seemingly lit up room. Just because everything and everyone is lit up about it does not mean the light is clean. If we look behind the shade what will we find? A light fixture covered in dust? Ick loves to stick to bulbs!

When I realized how much ick I had sticking to me, I still hung onto the wheel and kept driving, even after I knew Jesus was the solvent to the ick. I did not let Jesus take the wheel. He's the one guy we should let drive! I wanted Him to stay in the backseat with His seatbelt on and just pass me the goodies as I drove. Then I had a wet moment which opened my eyes to the fact that even though I knew all of this... all that I've shared with you... I was lost. GPS didn't help. I was surrounded by ick. I took God's free gift for granted. It may be free, but you still must count the cost. I didn't. Just like Jesus said in John 11 when He wept, "The days will come upon you when your enemies will build an embankment against you and encircle you and hem you in on every side." I was surrounded; I was in trouble.

"Then spake Jesus again unto them, saying, I am the light of the world: he that followeth me shall not walk in darkness, but shall have the light of life." John 8:12

To be like Jesus is to become a beam centered on God.

Our beam, like His, will determine the gravity of God's presence.

Are we balanced?

Just a Thought

Jesus is the Light of the world *and the crown of life*. Consider this:

We have already "evolved" into supreme beings when God created us into His likeness and image. We have since regressed into a corruptible subsistence of existence when our earliest ancestors decided to *come out* of the light of God. Jesus came to show us what it is to be human *in* the light of God. Jesus came to show us what it is to be human *in* the light of God as *sons and daughters of God.* We are already fashioned to dwell in the climax of evolution. We are in no need of genomic progress. We are as beautiful as we can be. All we need is His light to crown our perfection and heal our deformity and diseases caused by sin.

While pagan worship has praised the god of the sun since time immemorial, Jesus came as the one true Son! Within His light we are who we are meant to be and in Him we are abiding in the apex of existence. God is the summit of our trekking ambitions for vitality and to be immortal. All else pales beside it. Decomposes eventually. Outside of the light of God we are in danger of His *wHOLeY* powerful radioactive emissions. Inside the light of God, we are protected by His impenetrable membrane.

We cannot constitute in our greatest imaginations a being more glorious than God. We conceive and construct images which always falls short of the immaculate nature which is already God. The I AM. The I AM is *here and now* and we are born of His loins so to speak. Our only obstacle to become as He intended (in His likeness and image), as HE IS, is ourselves. We are one step away from the crown in the "evolutionary chain." A step which requires our midsection to bend rather than our legs to stride, or worse, our egos to plump. As we humble ourselves and fall to our knees we are lifted up and accepted in... into... onto the spire of the summit where God IS. To those who condemn creation as proud *Darwinians*, know this: to be like Jesus should be our proudest and zealous goal if we truly understood how close we already are to His sublime nature... not even an "era of evolution" away!

We are not gods in of ourselves, no matter how long we meditate, no matter how hard we *will* it. Nor if we conjure in laboratories a genetic solution to replace our futile anthropological quest for the holy grail. The true fountain of youth is in Jesus as He is (self-proclaimed) the *living water*. We are God's children (despite our own will to refuse to believe it), and this is proven mathematically in the written WORD and exemplified by Jesus as the living WORD. Humility is what we all require. We can do it by choice and come humbly before God, asking forgiveness, or we will be forced and humiliated afterward. Forced by the one unified force as it is unveiled to reveal who is on the inside of the light and who is outside of it. If not joined with God, we will be humiliated not because God will be a bully out to make fun of us, but because we will find ourselves on the wrong side of the force and know by our choice we alone were guilty of it. It will be humiliating. God agonizes over this and Jesus wept. So, let's choose humility before we are humiliated.

All of humanity since long ago has been chasing after the fountain of youth and today droves of people are driven to cosmetics, medicine, medical procedures and [rolling out the red carpet to] genetic manipulations to make us more beautiful. CRISPR CAS**9** (*nine*) technology is the beginning to a future world of gene editing. To become better, healthier humans… and to live longer. To live forever? To cheat death?

Chasing after manmade or "alien" formulas and elixirs are more far-reaching, literally and genetically than to rest in God's will and to let Him do the "reaching" for us. To arrive at an insurmountable and incorruptible genomic state of existence we are one man away. One man who walked among us two thousand years ago. An evolutionary climax is not generations ahead of us and the zenith to all our own efforts to better ourselves will prove to be a culmination of a concaved undertaking in the end. Whether we venture to reach physical perfection or spiritual nirvana by our own mediocre meditations coupled with medical facilitation, it will never be enough. All our brain power will never unify and never measure up to *barely enough*. Together as mankind pursues conscientious cohesion we will never muster up enough energy to prolong one fraction of a second of power needed to take one breath of air, much less keep us animated as living beings without the source. We need God. Oh, btw, don't settle for aliens! Don't succumb to biomechanical upgrades or alien DNA-altering procedures. When the "aliens" come they will bring with them their genetic elixir to "human evolution." It's in Satan's playbook.

Moreover, we will struggle as we entangle ourselves further (or is it farther, lol) into a web of technology as if it is incorruptible. The alluring lights of our flashing digital age is a snare. The lights are temporary! They will eventually turn off as the Power Breaker of the universe pulls the lever. *Trains, planes, and automobiles will crash and crush us. Power to our enviable tech will crash but not before the energy to that power crashes into our flesh with its invisible teeth of radioactive incisors.

We need God. If we surrender our soul to Him, He will give us His spirit. This is the only way to unlock us from our corruptible flesh, become *as the light* and safe from the external licking flames of hellfire, much less radiation. Claws. Teeth. Or anything harmful to us, for that matter!

God's way is non-corrosive and pure. He had a plan. The blueprint is still available. Big equations to build big machines is not moving forward. It's backward. Does bone-crushing concrete and flesh-piercing metals paint a picture of a future full of thriving life and sustenance? Nanotech implants with metals for electrical conductivity? Conductive to what? Truly, does it paint your utopian dreams? Does radioactive energy appear in your vision of a bright future or a dystopian society? How about a post-nuclear world? The blueprint of God's Kingdom for us has no such things.

God has a better way. Jesus, the light of the world, will be our sun. No more bulbs. No more LED's! Bioluminescence in everything may be our lightbulbs, but the glory of God's light, the Bible says, will replace the sun. He will brighten everything, so we can expect to see visible colors and hues from His spectrum we have never before seen... it will be glorious! Heavenly non-local teleportation will replace not just all modes of transportation, but communication too. No more crushing substances or harmful radiation. Appear and/ or make things appear by thought processes will be 2nd nature. We wouldn't miss technology because we would *be technology*. No more corrosive hardware because we would be sufficient software efficient for clean transactions. Why would we long to fly in planes or endeavor to fly in spaceships if we could fly ourselves within all dimensions in all universes? Why would we miss texting our friends if we could deliver messages in telepathic holograms or some kind of instantaneous delivery of information free from radio-frequency emissions? Those who have claimed to have visited Heaven in NDE's (near death experiences) have all said that the answers to all their questions were instantly downloaded into their minds as soon as they had thought of it. Coincidence? Man's or alien tech is not the future. Go God's way!

Our eternal hope for everlasting life with Him is God's dream come true! Yes, He has a dream… His dream is to *walk in the garden* with us. This was His dream from the get-go. For us to live a spiritual life as He intended from the beginning, free from corruption. His way is perfect. Perfectly pure. Perfectly sound. Perfectly bright. His love will endure forever, and His way is incorruptible. His promises are true. They've never been broken, and they never will be. We can look forward to an eternity of this with Him because of Jesus. The beam of His open arms will never fold, and the beam of His warm smile will never shut to those who love Him. The beam of Jesus rests on the unmoving justice and goodness of God the Father. Only in their union will we be in union with each other, balanced on the scales hanging from the beam of His perfect justice and insurmountable holiness. Made whole. Made pure. No more ick. Will we ever know how much it cost to pay the price for sin? God is absolutely loving and gracious, but He is absolutely just, as well. Justice without mercy would seal our doom, so, let's thank Him for using Himself (His begotten son) to pay the ransom for our filth, because we certainly would not have been able to pay the price to enter the cosmic carwash necessary to scrub us clean.

God is constant in His ways, but He also lends a different voice to various mouthpieces. Sometimes He has a soft-spoken word. A tender touch. Other times He will flex His muscles and have a word of judgement. Sometimes they come in seasons and other times they run parallel to each other. Grace, mercy, justice, and judgement are integral parts of His continuous love. But He is Holy Holy Holy above all else. God, as the source of One Force, is the only constant, and therefore, the only source trustworthy enough to depend on in this dying world. God is the same yesterday, today, and tomorrow. We may not see tomorrow, and Jesus invited us to "sup" with Him yesterday for every future day, so today is the absolute perfect day to RSVP. Let's accept His invitation and rest assured of tomorrow and of all the following morrows.

"But if we walk in the light, as he is in the light,
we have fellowship with one another,
and the blood of Jesus, his Son, purifies us from all sin."
1 John 1:7

"Therefore, come out from among them
and be separate, says the Lord.
Do not touch what is unclean, and I will receive you."
"I will be a Father to you,
and you shall be my sons and daughters
says the LORD Almighty."

2 Corinthians 6:17-18

Epilogue

W e're only halfway to the end. My original manuscript was too "beefy" according to my publisher, and so I decided to cut my book in half. I had already begun a 2nd book which will now be my third, but first, let's concentrate on the 2nd which was originally part of this first book. Lol! I think it was all meant to be because I am excited to name the 2nd half of this book God is the Boom, and now I am able to officially do that. I realize it is another cost to you (and me) to finish the story in another book, so I have discounted it as much as my publisher and retailers would allow. There is a limit, unfortunately. It was either that or having this book be text-size. For some of you, you may already have God is the Boom if you originally purchased the set of both books (Jesus is the Beam + God is the Boom) which I made available at a discount.

In the following book I will share what God so graciously revealed to me in January 2020 as I was writing the 2nd half of my manuscript. I was amazed and I'm still in awe of the whole thing. God's formula is so simple and the writing on the wall does not end with this book. The formula may be small, but the proponents are endless. I'm certain others will agree and begin their own summaries and reports of how the quantum cross... this *theory of everything*... pertains to their own line of work. For me, there are a few more accounts I must explain in regard to the equation of the cross.

It saddens me that I am not able to reveal it to you right now on this page. It's simple, yes, but my personal story... not so much. We still have to talk about the wet moment. How and when we could possibly walk through trees. We still haven't reached the clouds of angels or the fall-on-your-face moment of truth. In the next book we will know how good God is, by ripping off some Band-Aids covering wounds we didn't know we had. Wounds of pride and hatred caused by a scheming brood of vipers led by the ancient serpent himself, Satan. How will we know how good God is until we know how bad the devil is?

The Bible warned us that Satan will dominate the entire world and that he is a sneaky snake and a wolf in sheep's clothes, so it should not surprise us one bit to know unbelievable conspiracy theories are actually quite believable. And that whether or not we believe in God or Satan, the elite global leaders of this world are admittedly participating in ancient rituals clearly defined as Satanic by Biblical description. The push for globalization in our day and age proves we truly are living the "last days" as foretold in the book of Revelation. The devil is hard at work twisting good things into bad, and in my next book we will introduce many of his tactics. He's still handing out shiny red apples!

We will see how Satan is a lot like entropy, leaving us less than whole, as he uses God-created forces against us, and how the Electromagnetic Force is no different than the Weak and the Strong. We have underestimated the true Strong Force as we overlooked God the Father, God the Son, and God the Holy Spirit. We will see Gravity, the inscrutable of the Forces, as it operates with the EM Force finally being revealed for all to see atop the hill of Calvary.

God is the Boom will have a bang of an ending as I reveal the Holy Grail and the purpose of Jesus following quantum mechanical laws understandable to the quantum and non-quantum mechanical person. In addition, we will see how we are unified with the unified force in an unbreakable circuit. When I began writing my story over 20 years ago, I imagined my "reveal" of the *cross as two sticks of time igniting a flame of space* to be published in a little "Golden Book" for children because of the elementary nature of it. Also, I envisioned the little Golden Book to be a lampoon show-off to technical over-thinkers. I envisioned wisecrack scientists hitting their foreheads with the palms of their hands as the lightbulb turned on for them when they turned the colorful square page showcasing the formulation of the cross. At the time, I didn't know there was more to the formula other than time squared ($-^2 = +$). Now I see the serious implications of the force in all things. The one standard force! The Strong side of the force and its hellfire fury is one side of the coin, or rather, it is ¼ of the cross! It takes a big bite out of life. We will talk more about the electromagnetic (EM) side (quarter) of the force and how it is the most underestimated of the four and how its discharge of energy is measured by the quota of its waves and frequencies (not talking about lightning) and what it means to the health of each one of us, all lifeforms and our planet.

I realize much of my writing is conjecture not padded with experimental data or footnotes of proofs regarding the cross formulation. The cross truly speaks for itself and I do not feel pressured or required to confirm my claims with laboratory tests. I write this not with the intention of presenting my visions with experimental data because for one, I haven't access to a lab, and two, it is too much for me to handle on my own (it goes on and on and on). Three, because I know this is truly the theory of everything, I expect the experts in various fields to test what I offer and write their own theses. I am confident enough in everything I have presented to you to stimulate you, the reader, to explore what I've expounded. Whether you are a scientist or not, I encourage you to probe into all my claims and set an inquest for your own findings as you examine the details in the stitching of the fabric of your own blanket... *comforter.* Get out of your comfort zone! Analyze it. Scrutinize it. Do the groundwork necessary to till the hardened ground... the framework of your reality.

Fact-finding is not hard. Facts are everywhere above ground, but the devil buries truth so sometimes you must get out the auger and drill deep. The spiral is good for something! Just don't get swept away by it! The one thing Satan has not successfully buried is the Bible. Its truths are timeless and easy to read. Ask God to *open your eyes*.

My journey with Jesus is ongoing and I urge you to read the 2nd book without respite, so you do not lose momentum as my story continues to build. We will tread upward and crash downward on more harrowing adventures and come face to face with demons and meet angels ready for battle. I invite you to extend your reach and take another step. Do not interrupt your sojourn with me by stopping now, please. There's much more, and I saved the best for last. We have not yet arrived at the peak where God's formula will crescendo to a climax and reveal Jesus as never before seen. Jesus is truly the math-maker and the math-breaker. Jesus is the quantum solution of all matter (lost and found) and He is the altruistic missing link to mankind's search for THE WHO AM I's.

We think it's complicated. Life. It's not... it's actually quite simple. God is good, the devil is bad. More accurately, God is holy, and Satan is evil. All the signs are there all around us wherever we go... the writing is on the wall and it doesn't take a rocket scientist to read what it says. All our choices can be and should be made using this simple formula. God is good. Jesus is everything! Hugs and *kisses,* Gwyn

Bibliography

M ost references have been cited in-text within the manuscript unless noted below. All historical dates of or pertaining to events and names included in my manuscript were taken from or verified with Google.

1. Quote by Einstein, Albert. books.google.com... Renn, Jürgen (2007) The Genesis of Relativity: Sources and Interpretations. Dordrecht, The Netherlands: Springer, Page 124

2. Seife, Charles (2000) Zero: The Biography of a Number. Harmondsworth, Middlesex, England: Penguin Books, LTD. Page 57

Image Copyright Index

2. Standard Model, Standard_Model_of_Elementary_Particles.svg: MissMJderivative work: Alefisico / CC BY-SA (https://creativecommons.org/licenses/by-sa/3.0), https://commons.wikimedia.org/wiki/File:Standard_Model_of_Elementary_Particles-es.svg

3. Molecules, Alex Oakenman/Shutterstock.com

4. Atom, Quantum Mechanical Model, VectorMine/Shutterstock.com

5. 10 Commandments, AlexLMX/Shutterstock.com

6. Mammalian laminin reprinted from Journal of Molecular Biology, Vol 150, Copyright 1981, pp 97-120, Engel et al, "Shapes, Domain Organizations and Flexibility of Laminin and Fibronectin" with permission from Elsevier

7. Non-cross laminin sketch, original rendering by Gwynevere Lamb

8. Laminin Schematic, Maiaaspe / CC BY-SA (https://creativecommons.org/licenses/by-sa/3.0), https://commons.wikimedia.org/wiki/File:Schematic_Diagram_of_Laminin_111.jpg

9. Caduceus, Tribalium/Shutterstock.com

10. Laminin molecule diagram, original rendering by Gwynevere Lamb

11. Medieval Sword, Peyker/Shutterstock.com

12. Roman Numerals, Luisrftc/Shutterstock.com

13. Egyptian Numerals, Sidhe/Shutterstock.com

14. Mayan Numerals, Peter Hermes Furian/Shutterstock

15. Mayan Head Numerals, Sidhe/Shutterstock.com

16. Aztec Numerals, original rendering by Gwynevere Lamb

17. Sumerian Cuneiform Numerals, Sidhe/Shutterstock.com

18. Sumerian Cuneiform Tablet, David Herraez Calzada/Shutterstock.com

19. Nippur, Old Babylon, circa 1899, Morphart Creation/Shutterstock Greek Attic Numerals, Courtesy of Simon Ager, https://www.omniglot.com/writing/greek.htm

20. Greek Attic Numerals, Courtesy of Simon Ager, https://www.omniglot.com/writing/greek.htm

21. Greek vs Hebrew Numerals, Courtesy of Daniel, https://menorah-bible.jimdofree.com/english/structure-of-the-bible/alphabets-and-numerical-values/

22. Aramaic Numerals, Otfried Lieberknecht / CC BY-SA (https://creativecommons.org/licenses/by-sa/3.0), https://commons.wikimedia.org/wiki/File:Numeral_signs_aramaic.svg

23. Greek Syllabic Script, Courtesy of Simon Ager, https://www.omniglot.com/writing/greek.htm

24. Abacus, Pearson Scott Foresman / Public domain, https://commons.wikimedia.org/wiki/File:Abacus_(PSF).jpg

25. Chinese Modern Numerals, HAL-Guandu / CC BY-SA (https://creativecommons.org/licenses/by-sa/4.0), https://commons.wikimedia.org/wiki/File:Chinese_numerals_financial.png

26. Chinese Shang Numerals, Gisling / CC BY-SA (https://creativecommons.org/licenses/by-sa/3.0), https://commons.wikimedia.org/wiki/File:Shang_numerals.jpg

27. Chinese Rod Numerals, Courtesy of Simon Ager, https://www.omniglot.com/chinese/numerals.htm

28. Chinese Suzhou Numerals, Courtesy of Simon Ager, https://www.omniglot.com/chinese/numerals.htm

29. Mayan vs Babylonian zero, original rendering by Gwynevere Lamb

30. Samuray Studio/Shutterstock.com

31. Samuray Studio/Shutterstock.com

32. Brahmi Numerals, Otfried Lieberknecht / CC BY-SA (https://creativecommons.org/licenses/by-sa/3.0), https://commons.wikimedia.org/wiki/File:Brahmi_numeral_signs.svg

33. Cartesian Coordinates, KoenB / Public domain, https://commons.wikimedia.org/wiki/File:2D_Cartesian_Coordinates_nl.svg

34. 3D Cartesian Coordinates, H Padleckas, public domain, no alterations, https://commons.wikimedia.org/wiki/File:3D_Cartesian_coordinates.PNG

35. 3D Axes, original rendering by Gwynevere Lamb

36. Multi-dimensional Axes, Original rendering by Gwynevere Lamb

37. Fibonacci Spiral, LeonART/Shutterstock.com, altered: added numbers and numeral sequence

38. Parthenon floorplan, Argento / Public domain, https://commons.wikimedia.org/wiki/File:Parthenon-top-view.svg

39. Golden Ratio, Eisnel at English Wikipedia / Public domain, https://commons.wikimedia.org/wiki/File:Golden_ratio_line.png

40. 2D Graph of hyperbola. Courtesy of © J. Schauberger Verlag, Bad Ischl, Austria

41. 3D hyperbola. Courtesy of © J. Schauberger Verlag, Bad Ischl, Austria

42. 3D hyperbola. Courtesy of © J. Schauberger Verlag, Bad Ischl, Austria

43. Computer representation of the hyperbolic cone and spiral animations by Dr. Ines Rennert and Dr. Norbert Harthun © PKS, courtesy of J. Schauberger Verlag, Bad Ischl, Austria. Altered images by Gwynevere Lamb of inverted cone and spiral and adding 2nd flow to spiral.

44. Torus, Adith George 1840427 / CC BY-SA (https://creativecommons.org/licenses/by-sa/4.0), https://commons.wikimedia.org/wiki/File:Torus-.jpg

45. Vortex math. Original rendering by Gwynevere Lamb

46. Vertex quadratic functions, matma/Shutterstock.com; added arrows and definitions

47. Fibonacci Spiral, LeonART/Shutterstock.com. Inverted and added 9

48. Spiral, Andrew Fyfe / CC BY 3.0 NZ (https://creativecommons.org/licenses/by/3.0/nz/deed.en),

https://commons.wikimedia.org/wiki/File:NZ_flag_design_Koru_(Black)_by_Andrew_Fyfe.svg; cropped image of flag.

49. Spiral, Paper Wings/Shutterstock.com

50. Kelvin-Helmholtz instability, Stability of a vortex sheet roll-up, courtesy of Malek Abid and Alberto Verga, PHYSICS OF FLUIDS, VOLUME 14, NUMBER 11, 2002, DOI: 10.1063/1.1502660

51. Atom (orbital), Quantum Mechanical Model, VectorMine/Shutterstock.com; cropped.

52. Standard Model, Standard_Model_of_Elementary_Particles.svg: MissMJderivative work: Alefisico / CC BY-SA (https://creativecommons.org/licenses/by-sa/3.0), https://commons.wikimedia.org/wiki/File:Standard_Model_of_Elementary_Particles-es.svg

53. Proton and Neutron with Up and Down quarks, Harp / CC BY-SA (https://creativecommons.org/licenses/by-sa/2.5), https://commons.wikimedia.org/wiki/File:Proton_and_neutron.jpg

54. Proton and Neutron, original rendering by Gwynevere Lamb

55. Standard model of Elementary Particles Cush / Public domain, https://commons.wikimedia.org/wiki/File:Standard_Model_of_Elementary_Particles_Anti.svg

56. Gun-type assembly for Little Man and Fat Boy, Fastfission / Public domain, https://commons.wikimedia.org/wiki/File:Fission_bomb_assembly_methods.svg

57. Little Boy / Fat Man photos, ShanekPPS / CC BY-SA (https://creativecommons.org/licenses/by-sa/4.0), https://commons.wikimedia.org/wiki/File:Little_Boy%2BFat_Man.jpeg http://foxemerson.com/the-spiral/

58. Caduceus, Tribalium/Shutterstock.com

59. W.H.O. logo, FIHIHF2013 / CC BY-SA (https://creativecommons.org/licenses/by-sa/3.0), https://commons.wikimedia.org/wiki/File:Who-logo.jpg

60. Caduceus, Rama / Public domain, https://commons.wikimedia.org/wiki/File:Caduceus_large.jpg

61. Baphomet, Eliphas Levi / Public domain, https://commons.wikimedia.org/wiki/File:Baphomet.png

62. Genesis 5 Genealogy Table. Original rendering by Gwynevere Lamb

Index

CPSIA information can be obtained
at www.ICGtesting.com
Printed in the USA
LVHW080823220421
684433LV00009B/68/J